Major Peter Ve
1953 until he re
he served with
Cyprus and Ade
Eton and Trinit
book, the first being *The Standard Bearer*.

THE MICKS

The story of the Irish Guards

PETER VERNEY

Foreword by

Field-Marshal Sir Gerald Templer

KG, GCB, GCMG, KBE, DSO

ABRIDGED EDITION

PAN BOOKS LTD: LONDON

First published 1970 by Peter Davies Ltd
This abridged edition published 1973 by
Pan Books Ltd, 33 Tothill Street, London SW1

ISBN 0 330 23632 6

Printed in Great Britain by
Richard Clay (The Chaucer Press), Ltd, Bungay, Suffolk

This book is dedicated to all
Irish Guardsmen, past, present
and future

The Irish Guards

by

RUDYARD KIPLING

We're not so old in the Army List,
  But we're not so young at our trade,
For we had the honour at Fontenoy
  Of meeting the Guards Brigade.
'Twas Lally, Dillon, Bulkeley, Clare,
  And Lee that led us then,
And after a hundred and seventy years
  We're fighting for France again!

Old Days! The wild geese are flighting
  Head to the storm as they faced it before!
For where there are Irish there's bound to be fighting,
  And when there's no fighting, it's Ireland no more!
          Ireland no more!

We're not so old in the Army List,
  But we're not so new in the ring,
For we carried our packs with Marshal Saxe
  When Louis was our King.
But Douglas Haig's our Marshal now
  And we're King George's men,
And after one hundred and seventy years
  We're fighting for France again!

Ah, France! And did we stand by you,
  When life was made splendid with gifts and rewards?
Ah, France! And will we deny you
  In the hour of your agony, Mother of Swords?
Old Days! The wild geese are flighting,
  Head to the storm as they faced it before,
For where there are Irish there's loving and fighting,
  And when we stop either, it's Ireland no more!
          Ireland no more!

Contents

List of Illustrations

Except where otherwise stated, photographs are reproduced by permission of the Irish Guards

Foreword

by

FIELD-MARSHAL SIR GERALD TEMPLER, KG

Compared to many regiments of the British Army, the Irish Guards form the subject of few volumes for the sole reason that they are 'not so old in the Army List' – but what a name they have achieved in those few years. There is first of all *The Irish Guards in the Great War* by Rudyard Kipling – whose only son, with whom I was at school, was killed in the Regiment. With his great love of British soldiers and his affection for the Irishman in particular, Kipling put his heart and soul into that splendid record. He made no attempt to comment on strategy and tactics, but what he recorded, with such infinite love and labour, was the spirit imbuing a regiment which was only fourteen years old at the outbreak of the war. In that short span of years it had indeed become not only a first-class regiment, judged even by the superlative standards of the Brigade of Guards, but also a happy band of brothers, with all the military virtues that those words stand for both in peace and war.

And then there is the *History of the Irish Guards in the Second World War* written by Major Desmond FitzGerald and published in 1949, which has always struck me as a model history of this type.

The foreword to the second of these books was written by Field-Marshal Earl Alexander of Tunis – 'Alex' to all members of the Regiment past and present. I know that all Irish Guardsmen will wish me to say something about that remarkable man, since he was the ideal of the many virtues which form the basis of regimental and Army, and indeed Christian life. I want to do so because I feel an interloper in contributing to this regimental story, in the writing of which he was so intensely interested. It was, of course, the wish of everyone concerned that he would do what has now fallen on me, but it was not to be.

He was indeed my ideal of a *beau sabreur.* I loved the man as if I had been in the Regiment myself. I knew him for thirty-five years and first came under his spell when he was Chief of Staff Northern Command in York, where I found myself as his number two. He was kindness itself to me; he made me work – and encouraged me to play. Perhaps he took a remotely cousinly interest in me too because I had been born and bred an Irish Fusilier and had roots in County Armagh, only five miles from his old home at Caledon in County Tyrone. I think he relished the fact that a forbear of mine had commanded that fine old Irish fighting regiment, the Connaught Rangers, in the latter stages of the Peninsular War.

Years afterwards I was privileged, as the Divisional Commander, to attend the parade of the 1st Battalion of his old Regiment when Lance-Corporal Kenneally received his Victoria Cross for his splendid deeds at the famous Bou Arada – the 'Bou' as it was known to all fighting soldiers in North Africa at that time. It moved the comparatively few spectators as much as if we had been Micks ourselves. How proud we were to be present, and to be able to share in Alex's pride and admiration, not only for the recipient but for all the officers and men who had taken part in that epic fight.

Perhaps these few memories are not entirely inappropriate here, for the reason that this great soldier had received his early training in his beloved Regiment. He would, I am sure, have said at once that it was his old Regiment which had largely made him what he was. A great family Regiment, and how much that means can only be appreciated by someone who has had the honour to serve in one. (As recently as the 1st Battalion's operations in Aden in 1967 the fathers of three of the four company commanders had served in the Irish Guards in their time.)

I have two personal memories of the Regiment which I

shall always cherish. To me they seem to epitomize the spirit among all ranks of the 1st Battalion – the only Battalion of the Irish Guards which I had the privilege to know in the Second World War. I am sure these memories are applicable to the whole Regiment. The first, a visit in Tunisia to an olive grove, interspersed with fig trees. The Micks, as usual and quite rightly, had made themselves as comfortable as possible. Tied on to one of those trees was a home-made Irish Republican flag – and behind peered a snub-nosed Mick face – in an attempt to sense my reactions. There were none – except of amused admiration.

And then at about the same period, towards the end of a flogging three-day company training march (such junketings could never be exactly popular with battle-tried units) along came Number 3 company commanded by that splendid young officer, John Kennedy – killed later in Europe. In spite of all my precautions he spotted me a couple of hundred yards away, and out he ran beside his Company – tired indeed as they were – and pulled them together and marched them to attention with the Company piper playing what used once to be considered that good old rebel tune, *The Boys of Wexford*. I can see them now – every Guardsman marching with a slight flick of the wrist in that quite unauthorized (I am sure) manner which, for me, demonstrates the panache of the Irish Guards.

I like to think that this book commemorates not only the story of the Irish Guards, but through them the services of all Irishmen who have served in the British Army – and in particular the officers and men of the five disbanded Irish regiments – and even perhaps those other Irish mercenaries of old who fought as the Wild Geese over the water.

This new volume is not a regimental history in the strict sense, but it does portray in the most vivid fashion the spirit of a family regiment. Reading it, one can sense something of the magic which all who have served in or even known this great Regiment have felt.

Introduction

This is a personal story: the story of a regiment over the first seventy years of its existence. An anonymous Irish Guards private in the Great War said, 'We was like fleas in a blanket, seein' no more than the next nearest wrinkle' – it is from that very narrow viewpoint that this book has been written.

The inherent risk in such an approach is that the author cannot give due recognition to the part played by others in the events he describes; and from time to time readers of this book may receive the impression that the Irish Guards were fighting in limbo and without support. This was, of course, rarely the case.

Yet there is some justification for this treatment, since in battle the fighting man knows remarkably little of what is going on outside his immediate compass. He may hear of the doings of other regiments or brigades, but the reports he receives will be both stale and garbled; of more concern is that he should get his breakfast, have confidence in his officers, and, most important of all, be able to associate himself with that indefinable spirit of his regiment, which is known as morale. It is this spirit which sustains him in battles and in boredom – and if there is one thing which the Irish Guards have always possessed in full measure, it is spirit.

The Irish Guards have always been basically an Irish regiment. At the outset of a major war the Irish percentage is almost complete, and as the war progresses the proportion shrinks; but it is a characteristic of Irish regiments to be able to absorb outsiders and turn them into the national product. The words of a sergeant serving in the 1st Battalion Irish Guards at Anzio in 1943, 'I may not come from Ireland, but I am as fine a — Mick as you are', typify the binding family tradition of the Irish Guards.

There are many facets to the Irish character. The strong religious feeling, the intense loyalty and overwhelming exuberance in all the Irishman does on the field of battle are very evident in this book. Not for him the dogged dourness in defence of the average British soldier – this is quite foreign to his nature – and although he will absorb punishment with a resigned fatalism, he is no stoic, and such an attitude is totally unsuited to his temperament and inclinations. The Irishman is *par excellence* an aggressive soldier; he needs leaders who know him and his peculiar ways, but above all he needs a challenge: only this will bring out the best in him.

This is as true with the Irish Guards as with any Irish regiment; but what makes the Micks a unique institution is that, imposed upon their common Irish birthright, are the traditions, training and discipline of the Brigade of Guards. And it is within this framework that the Irish Guards have existed for the last seventy years.

I have been especially fortunate in being able to use Kipling's book on the Great War so fully, and for this I am deeply indebted to Mrs Elsie Bambridge, Rudyard Kipling's daughter and herself the widow of an Irish Guards officer. My other principal source has been Major Desmond FitzGerald's *History of the Irish Guards in the Second World War*. This has proved of immense value and I am most grateful to the author for his permission to use it so extensively. I have also generously exploited Brigadier J. O. E. Vandeleur's recent book *A Soldier's Story*, and I am most conscious of the debt I owe him.

I am particularly grateful to Sir Charles Petrie for his invaluable advice on the form and content of this book and for his constant encouragement; to Colonel Tris Grayson, who largely compiled the chapter covering the period between the two World Wars; and to General Charles Haydon, who from his personal knowledge virtually rewrote the chapter on the Hook of Holland and

Boulogne adventures; to Brigadier Denis FitzGerald, who had the unenviable task of reading the early drafts and whose advice was quite invaluable; to Captain Christopher Wolverson, whose time-consuming and dusty research into the early days of the Regiment was so helpful; to the staff of Regimental Headquarters and particularly Major Tony Mainwaring-Burton, for a lot of painstaking work over the past three years; and to Major David Faulkner, who was the true instigator of the idea for this book in the first place. And above all to the many Micks and ex-Micks, who prefer to remain anonymous, but without whose help and advice this story of their Regiment could never have been written.

May 1970 PETER VERNEY

Chapter 1

Formation*

As Queen Victoria's long reign drew to its close, she wanted nothing but to be at peace with all her peoples. She had never really understood the Irish, classing them as a troublesome breed; and there is little doubt that she had been badly frightened at the time of the Fenian uprising in the late 1860s. She had only made two trips to Ireland in her life, and they had been many years before. Yet here were these same subjects fighting with breathless bravery in the South African War, for a cause which could hardly be called their own.

Victory in that war seemed as remote as ever. For the first time for many decades British troops were suffering defeat. December 1899 had been one long chapter of humiliation – the reverse at Magersfontein; the débâcle of the retreat from Stormberg; the resounding failure at Colenso. The only feature from which those at home could draw any comfort was the undoubted gallantry of their soldiers – and none more gallant than the Irish regiments which had suffered frightful casualties.

The arrival of Lord Roberts as Commander-in-Chief brought a steady improvement in British fortunes in South Africa. Within three weeks of his arrival the ubiquitous Boer commander Cronje, long a thorn in the flesh, had been out-generalled and forced to surrender at Paardeberg.

* The extracts from telegrams and letters used in this chapter are from the Private Papers of Queen Victoria in the Royal Archives and are reproduced here by gracious permission of Her Majesty the Queen.

Ladysmith was next, and the fighting to relieve that town was some of the bitterest on record. Again the Irish regiments greatly distinguished themselves, and suffered terrible casualties. So much so that on February 27th, 1900, the Queen herself was moved to telegraph to Sir Redvers Buller, '...I have heard with the deepest concern of the heavy losses sustained by my brave Irish soldiers. I desire to express my sympathy and my admiration of the splendid fighting qualities which they have exhibited throughout these trying operations. VRI.'

It is probable that there had long been a movement to create a Regiment of Irish Guards – and it is equally probable that the Commander-in-Chief, Lord Wolseley, was one of its principal advocates. This is pure conjecture, but it could be that observing the Queen's gradual change of heart he judged that the moment might be ripe to put forward, but subtly, his design. Be that as it may, some days before the Queen's telegram to Buller, he had written to Sir Arthur Bigge, the Queen's Principal Private Secretary:

> My dear Bigge,
>
> I am Colonel of the Royal Irish, once the Irish Guards, and I know from old acquaintance with it how much Irish soldiers appreciate praise, and how very highly they would value any expression of the Queen's approval. It would be a most gracious step on the part of the Queen to express herself satisfied with the daring and devotion to her Throne of her Irish soldiers now before the enemy in South Africa . . .
>
> Very sincerely yours,
>
> WOLSELEY

and, on March 1st, followed this letter to the Queen herself:

> Lord Wolseley presents his humble duty to the Queen, and ventures to congratulate Her Majesty upon the success of her soldiers in relieving Ladysmith. Lord

Wolseley feels that the Queen's appreciation of the gallant services rendered by her Irish soldiers will have a magical effect upon that sentimental and imaginative race all over the world.

Acting upon the Queen's initiative in this matter, Lord Wolseley would presume to suggest that the Queen should now order all her Irish regiments to wear the 'shamrock' in their head-dress on the 17th instant, to be worn by all her Irish regiments in future years on Saint Patrick's Day as a mark of Her Majesty's appreciation of the daring displayed by her loyal Irish soldiers in the recent operations near Ladysmith.

The Queen duly authorized the wearing of the shamrock on St Patrick's Day. But a greater mark of appreciation of the bravery of the Irish troops in the South African War was pending, for also on March 1st, 1900, a letter appeared in *The Times*:

To the Editor of *The Times*:
Sir,

May I venture to suggest, through you, to the authorities within whose province it may come, that now is a most opportune time to recognize the distinguished valour of our Irish soldiers who, in the Inniskilling Fusiliers, the Dublin Fusiliers and the Connaught Rangers, have shown to the world such conspicuous bravery in the many recent battles which they have fought with such brilliant dash and daring throughout our South Africa War. Is there not one mark of distinction and honour that can be conferred upon them and their country which belongs to Scotchmen and Englishmen, but is withheld from them? There are Scotch Guards and English Guards – why not add to the roll of glory a regiment of Irish Guards?

I am, Sir,
Your obedient Servant,
CUMMING MACDONA
House of Commons, February 28th, 1900

The next day Sir J. W. Maclure, MP, put down a question in the House of Commons suggesting a proposal to be submitted to Her Majesty to 'allow a regiment to bear the title of "The Royal Irish Guards" and to confer the first colonelcy on Lord Roberts'.

On receiving notice of the question Lord Lansdowne, then Secretary of State for War, wrote to Sir Arthur Bigge '... The fine conduct of the Irish Regiments during the war, and Her Majesty's gracious interest in them suggested to several of us the idea that the time might have come when the Queen might be pleased to authorize the creation of a Regiment of Irish Guards...' He added that he had intended approaching the Queen on the subject but had been anticipated by Maclure's question. Lord Lansdowne therefore sought the Queen's permission to answer the question to the effect that such a proposal was to be submitted and might be expected to receive favourable consideration.

Sir Arthur Bigge replied on March 5th: '... By a curious coincidence the Queen has during the past week been seriously considering the question of a Regiment of Irish Guards, thinking that the present was an opportunity for its creation. Therefore I am glad to be able to tell you that the Queen entirely approves of the idea and of your proposed reply to be given to Sir J. W. Maclure's question in the House of Commons on Tuesday next.' Furthermore, she had already asked the Duke of Connaught to speak to the Commander-in-Chief on the subject.

From then events moved with impressive rapidity. For, rather less than one month after the first suggestion to form the Regiment, the Irish Guards were created. Army Order No 77 dated April 1st, 1900, read: 'Her Majesty the Queen having deemed it desirable to commemorate the bravery shown by the Irish Regiments during the operations in South Africa in the years 1899–1900 has been graciously pleased that an Irish Regiment of Foot Guards be formed to be designated the "Irish Guards".' And so the Irish Guards came into official being.

By terms of the resulting recruiting orders volunteers were asked for from Irishmen serving in the other regiments of the Brigade of Guards, and later from those in other regiments. Simultaneously recruiting was started in Ireland and certain centres in England.

In those days the full establishment of a battalion was eight companies. That of the Irish Guards was an initial two; this was enlarged to three a few weeks later, and so immediate was the response to join the new Regiment that it was soon extended to four. At the same time the establishment of the Regimental Band was agreed. The first Director of Music, chosen from a hundred applicants, was Lieutenant C. H. Hassell.

It was no surprise that Lord Roberts was chosen as the first Colonel of the Irish Guards – for no greater Irishman lived. And he promptly ordered his two ADCs, Lords Settrington and Herbert Scott to transfer. At that stage the style of uniform and accoutrement had not even been discussed, so to mark their transfer the two officers devised a distinctive hat plume which Lady Settrington dyed in ink, to a blue-green as near the colour of the sash of the Order of St Patrick as she could achieve. One of these plumes still exists at Regimental Headquarters Irish Guards.

The first Commanding Officer was Lieutenant-Colonel R. J. Cooper, transferred from the Grenadier Guards. His Second-in-Command was Major D. J. Hamilton from the Royal Fusiliers, one time a Coldstreamer. The first intake of officers comprised Major G. C. Nugent, Lieutenant the Earl of Kerry, and Lieutenant and Adjutant H. Crichton from the Grenadiers; Lieutenant Lord Oxmantown from the Coldstreamers and Major Seymour Vandeleur, DSO, from the Scots Guards.

With their own traditions firmly implanted over the centuries and their reputations enhanced by fresh deeds of glory in the South African War, the other regiments of the Brigade of Guards looked upon the new arrivals with a benevolent, if somewhat quizzical eye. But had it not been for the generous support of the other regiments of the

Brigade throughout the early years of the Irish Guards, the latter would have found the task of establishing themselves incomparably harder.

No regiment in the British Army had been created from scratch for many years. They had no uniform, no customs, no traditions. Their officers had come from a variety of regiments with a variety of customs; from this mélange had to be created something similar to the older regiments of the Brigade of Guards, yet somehow subtly distinctive and unique to the Irish Guards. Their only bond at this early stage in their communal lives was that they were all Irish and intensely conscious of the honour shown to the whole Irish nation by the creation of their Regiment.

The recruiting scheme was prospering. The first actual recruit, and not a transfer, was one James O'Brien from Limerick; he was given the Regimental Number 7. Colour-Sergeant Conroy, from the Munsters, was Regimental Number 1.

On May 24th, 1900, Regimental Sergeant-Major J. Fowles was appointed Quartermaster to the budding regiment. To 'Ginger' Fowles, as he was always known, must go a great deal of the credit for the smooth evolution of the Irish Guards. He himself achieved a unique distinction for, some days after his new promotion and appointment had taken place, he appeared at the Queen's Birthday Parade in his previous role as RSM of the 1st Battalion Grenadier Guards providing the Escort to the Colour.

Fowles was a disciplinarian from the toe caps of his shining boots to the tips of his large pointed moustache. On one occasion he was guest of honour at a dinner one of the companies gave to their Company Commander who was about to get married. In the middle of the feast one of the men happened to pass the remark that the potatoes seemed a little underdone; whereupon Fowles promptly placed his host in close arrest and had him marched off to the guard-room for making a frivolous complaint. But Fowles was all gold at heart, and there was nothing he would not do for his Micks. They knew this and excused him his fierce and forbidding exterior, for he was a 'character' and Irishmen can forgive a 'character' almost anything.

Fowles was an enthusiast in all he did. He was also no inconsiderable performer in the boxing ring himself, and became the real instigator of boxing in the Army. Under his brilliant guidance the Irish Guards were soon an acknowledged force in the boxing world of the 1900s. Such encouragement was needed for a new regiment, and in this and many other fields Fowles provided it.

The form of dress for the new Regiment was the subject of much speculation, and the Press took the opportunity to proffer a number of suggestions – that they should dress like the Papal Guard in Rome; or wear green kilts; or, most horrible of all, wear orange tunics. There was, of course, no doubt that the Irish Guards would wear the tunic and bearskin cap of the Brigade of Guards, with an appropriate decoration.

All that now remained to be resolved was the regimental emblem. The star of St Patrick was chosen. The true star of St Patrick is slightly elongated. To avoid confusion with the star worn by the Coldstream Guards it was decided to adopt a squarer pattern, and it is this modified form of the Patrick that is now worn.

By September the Regiment mustered over 400 rank and file, and on the 20th of that month the training companies of the Irish Guards were amalgamated and the 1st Battalion Irish Guards was officially constituted. At the same time a Regimental Orderly Room was established and the first Lieutenant-Colonel Commanding the Regimental District was appointed – this was Colonel V. J. Dawson, CVO, from the Coldstream Guards. Colonel Dawson had been the first commanding officer of the 3rd Battalion Coldstream Guards on their formation in 1897, so he was uniquely aware of the problems attendant on the raising of a new battalion. Certainly the Irish Guards owe a great debt of gratitude to the understanding and quiet authority of their first Regimental Lieutenant-Colonel.

The Irish Guards were now deemed ready to take their

part in the ceremonial round in London. The majority of the battalions of the Household Brigade were still serving in South Africa, so troops for Public Duties were in short supply. The specially formed Royal Guards Reserve Regiment of reserve officers and men had borne the brunt – on one Guard ceremony the service of the three officers on parade, which was all they could muster, totalled an amazing fifty-four years, and this was by no means uncommon. Thus any assistance was more than welcome.

On March 3rd, 1901, the Irish Guards mounted their first Guard, under command of Major G. C. Nugent – but without Colours as they had still to receive them. By all accounts this Guard was not a great success, for the men presented a motley appearance as the uniforms were by no means complete, and certain items of ceremonial equipment were still awaiting issue.

A fortnight later came St Patrick's Day. Donning their national emblem the battalion, six companies of close on 700 men by this time, paraded at Church Service. After the service and while still awaiting to be dismissed, an orderly of the Life Guards, bearing a number of boxes, approached Colonel Cooper and handed him a letter. This was a personal message for the Colonel of the Regiment from Queen Alexandra, who requested that the shamrock in the accompanying boxes be distributed to the men of the Irish Guards as a gift from Her Majesty. Immediately the sprigs worn by the men were replaced by the royal gift. This spontaneous gesture on the part of the Queen was deeply appreciated and thus started a custom which is continued to this day. Each year until her death Queen Alexandra donated the shamrock, then Princess Mary, later the Princess Royal, undertook to make the annual gift. When she died in 1965 Queen Elizabeth the Queen Mother graciously consented to preserve the tradition which sets the seal on the St Patrick's Day parade. Each year it repeats its own unique brand of family intimacy and informal formality – and not a little poignancy.

The year 1901 was one of consolidation in other ways. In

April the King approved the choice of the regimental marches – *St Patrick's Day* for the quick march; and *Come Back to Erin* for the slow – later changed to the present *Let Erin Remember*, composed by Sir Charles Stanford. Shortly afterwards the first inspection of the new regiment was carried out by the General Officer Commanding the London District – the Major-General.

Meanwhile in South Africa the war had taken a fresh turn, for the Boers had now split into small commandos. These, unencumbered by baggage and living off the country, roamed at will harassing small parties, disrupting communications, and attacking isolated garrisons. This guerilla-type warfare persisted throughout the early part of 1901. To combat it, by summer the blockhouse system of lines of strongly constructed stone buildings, each manned by some eight to ten men and strengthened by ditches and wire, had been perfected. Alternate blockhouses were garrisoned by white troops, the remainder by coloured Cape Irregulars – known as the 'Black Watch' and notoriously trigger-happy.

To act as links between these fortifications and to dominate the surrounding no-man's-land required mobile troops, and among other mounted infantry reinforcements was raised the First Guards Mounted Infantry Company.

On October 2nd, 1901, 142 men drawn from the four regiments of the Brigade left London for Aldershot. In command of the Irish Guards section of the company was Lieutenant Lord Herbert Montagu-Douglas-Scott, DSO, the first of three generations of his family to serve in the Irish Guards.

After a hectic six weeks' mounted infantry training the 'Aldershot Mounted Foot' – as they were irreverently known – was passed fit for war service, and at the end of November sailed for South Africa. It was the Irish Guards section in the 1st, and later the 2nd Guards Mounted Infantry Companies, who were the first other ranks in the Irish Guards to see active service, and to suffer casualties in battle. Colour-Sergeant Hudson of the 1st Guards Mounted Infantry Company was the first Irish Guards non-commissioned officer to receive a medal for gallantry in action.

Before peace was signed in South Africa in early May the presentation of the first Colours to the Regiment took place. And on May 30th the new Colour was trooped at the King's Birthday Parade.

In August 1902 occurred a regimental occasion of a different sort. As the *Military Mail* recorded – 'So far the Irish Guards have managed to exist without a pet, though some have accused the custodians of the Tower of London of a fondness for sea-gulls. It was a state of things, though, which could not last, and the Irish Wolfhound Club has recently offered to make the Irish Guards a novel and appropriate presentation in the shape of a young Irish wolfhound as a regimental dog, and the proposed gift has been accepted on behalf of the Regiment by Lieutenant-Colonel R. J. Cooper, the Officer Commanding.' Eight entrants duly paraded at the Crystal Palace in October. The winner, thereafter known as 'Brian Boru' and given the Regimental Number 1463, became the first of a continuing line of Irish wolfhound regimental mascots.

Later that month the troops of the Household Brigade who had recently returned from South Africa took part in a Royal Review. A small contingent of the Irish Guards was formed up at the rear of the parade – the two sections of the Mounted Infantry Companies and a few transfers from the other regiments. At the head of this small body of men Lord Roberts marched past his Sovereign. Then, led by the Regimental Band, they marched to Wellington Barracks and were addressed by their Colonel for the first time. It was a proud moment.

Chapter 2

Early days

Recruiting was proceeding well. In March 1903 it was discontinued in the English districts; thereafter the flow of men from Ireland was enough to keep them up to strength. The bounty for a recruiting sergeant in those days was worked on a sliding scale – five shillings for a recruit for the Brigade of Guards, no less than thirty shillings for one for the Household Cavalry. And the Irish recruit, after satisfying the recruiter that his Irish bona fides was impeccable, set out for the Guards Depot at Caterham, home of the Brigade of Guards until 1960.

The Guards Depot course in the days before the Great War was of sixteen weeks' duration. Drill, musketry and physical training occupied all the recruits' time; there was no field training as we know it today. Evenings were devoted to 'shining parades' – when buff equipment was whitened, rifles were cleaned, and greatcoats folded. Attic rooms or outhouses were set apart for the pipe-claying of the white jackets which were worn as the normal training and everyday dress.

There were no messrooms. Food was collected from the cookhouse, carried to and eaten in the barrack-rooms. No leave was allowed for the sixteen-week period, but the recruit was permitted to go outside barracks over the weekends after tea each day, provided he could pass inspection from the Sergeant of the Guard – and this was sometimes no easy matter, for moustaches were obligatory in those days and Pomade Hongrois, however liberally applied, is no real substitute for hair on the upper lip. The 'rooker' or recruit walked out in the white jacket, thus being easily distinguishable from the 'qualified soldier' who wore his red tunic.

Pay for the recruit was ninepence a day. From this was deducted the cost of all cleaning materials and a small charge for haircuts, laundry and barrack damages. This left little enough; but these were the days when three halfpence paid for a pint of beer, and threepence could buy an ounce of tobacco or twenty cigarettes.

At the end of sixteen weeks the trained recruit became a fully paid private, and in the case of the Irish Guards, was then sent to the 1st Battalion. On arrival he was allocated to a company, according to his height. In those days each of the eight companies of the Guards and Infantry battalions contained four sections some twenty strong. There was no Headquarters Company as we know it today, the essential transport, signallers, clerks and other employed men being split between the rifle companies. Each company had a Colour-Sergeant – known as a Pay-Sergeant in the Brigade of Guards – who doubled in the roles now filled by the Company Sergeant-Major and Company Quartermaster-Sergeant.

Even within the battalion there was no centralized messing – and as at the Guards Depot food was eaten in the barrack-rooms. The Army Service Corps provided half a pound of bread – 'rooti' they called it – and three-quarters of a pound of meat a day for each man. This was supplemented by a messing allowance of sixpence a day for groceries and vegetables. As the company messing account was kept by the Pay-Sergeant of each company, and this worthy was often enough no great hand at figures, the standards of food varied considerably.

There were two canteens – the 'wet' and the 'dry' – as well as a regimental recreation room with draughts, chess and similar games. The running of these canteens was sometimes done by the battalion themselves, or else let out to contract.

Discipline was hard, but basically fair. The principal

crime was drunkenness. But the Spartan existence and, to us, severe discipline were accepted as normal and taken for granted. Morale seems to have been invariably high and the 'mucking-in spirit' at its best. The red tunic was worn on all occasions with the green forage cap. The soldiers looked smart and were smart, and after all they were getting an extra penny-a-day on the statutory shilling for being in the Brigade of Guards.

Promotion was comparatively quick, and for the serious soldier who wanted to better himself there were education classes, paid for by the officers and run on contract – for this was before the days of the Royal Army Education Corps.

There was little organized recreation. Every evening in the weekends, and on quite a few others besides, impromptu concerts were arranged in the canteen. Some old soldier probably acted as master of ceremonies and the talent in step-dancing and singing came from those around. From time to time family outings were arranged by the Quartermaster, Captain Fowles.

It was Fowles, too, who founded the 'Shamrock Minstrel Troupe' which features large in the battalion concerts of this period. In the Sergeants' Mess quadrille parties were all the rage and, for variation, 'ping-pong'. The Army Temperance Society was in its heyday and the local branches organized concerts and 'dry' entertainments. Later was laid the foundation stone of the Union Jack Club which still produces cheap meals and accommodation for the Services in London – then a revolutionary step towards the soldiers' welfare.

The families of the married men were a constant problem. The days were long since gone when families had to sleep together in barrack-rooms behind flimsy curtains, and now married quarters were provided in all garrison centres. To keep the wives employed and out of mischief they were contracted to do the battalion's laundry – with mixed success if the numerous complaints are anything to go by.

Within the Irish Guards one annual event stood out

above all others – St Patrick's Day. Activities started with the Drums marching round the barracks at reveille playing *St Patrick's Day, Brian Boru and* other Irish martial airs. At a morning parade the Royal shamrock was then presented, and at the conclusion of the ceremony the whole battalion marched to Mass, to return to gargantuan dinners. The afternoon was reserved for sports of all kinds, Gaelic football and hurley. When that was over another meal was produced followed by a concert which went on late into the night. The battalion was allowed a late start the next day.

Their world revolved around ceremonial duties – 'Public Duties' as they were known. Prior to 1903 the King's Guard had always mounted in the Friary Court of St James's Palace, but in that year King Edward decreed that whenever the King or Queen was in residence at Buckingham Palace the King's Guard of the Household Infantry should mount in the forecourt of Buckingham Palace. Except for a few minor developments and a reduction in the numbers on parade, the style in use today is precisely the same.

Another and always popular Guard, for it entailed a bonus of one and threepence a man and appropriate sums to the non-commissioned officers, was that mounted on the Bank of England. In addition there were one or two other smaller Guards – but without compensation.

In the summer of 1903 the 1st Battalion Irish Guards, who were then temporarily established at Pirbright Camp, were asked to provide a contingent for the state visit of the King and Queen to Ireland.

On July 17th the party, some 500 strong, arrived at North Wall, Dublin, for the first visit of the Regiment to Ireland. Everywhere the welcome was overwhelming. Their route to barracks was lined by the Royal Irish Constabulary, but they were quite unable to hold back the huge, cheering crowds who were soon marching along in the ranks, the girls clinging to the butts of the rifles.

They were in Ireland for ten days: lining the Royal route, or providing Guards of Honour at Viceregal Lodge in Dublin, or at St Patrick's College, Maynooth – where the King was delighted to see pictures of his two horses Diamond Jubilee and Ambush II dominating the hallway – and at Amiens Street Station on the departure of the Royal party for Belfast.

For the battalion, the highlight of the visit was the Royal Review in Fifteen Acres in Phoenix Park. Ten thousand troops and three Field-Marshals were on parade, and the feature of a wonderful day for the huge audience was the sight of Lord Roberts, 'Bobs' himself – the hero of All-Ireland – galloping out to take his place at the head of the Irish Guards contingent led by 'Brian Boru' and the Regimental Band.

One serious regimental omission was put right in 1905, for it was then that the Officers' Regimental Dining Club – the Quis Separabit Club – was formed. The inaugural dinner of the Club took place in the summer of that year at Willis's Rooms in King Street, St James's, and was deemed a great success. Particularly by Fowles, the Quartermaster, who was afterwards seen leaning against a lamp-post outside the restaurant gravely serving a poached egg from a huge plate to each passer-by.

For the next years the battalion was stationed in London, either at Chelsea Barracks or the Tower of London.

Barrack routine was unexciting and was varied, but not very often, by a Commanding Officer's Parade or by a drill parade in Hyde Park, the men in their very smart white jackets, and the officers wearing blue frock-coats. The only training was in musketry; this took place at Purfleet, near Southend-on-Sea – or when the battalion was stationed at the Tower, in the Moat – a damp and depressing place. Every year two weeks were spent on the rifle ranges at Pirbright.

Pirbright Camp in the early days of the century was nothing but a collection of tents. Cooking was done in field-kitchens on the banks of the Basingstoke Canal, and from there the food was brought in dixies to the mess-tins placed outside tents. Washing and latrine facilities were as primitive; and the canal was used for everything with little discrimination. Here, too, were washed and whitened the white jackets which with the 'Brodrick Cap' – like a muffin-man's they said – was the dress for musketry and field training.

In 1910 the 1st Battalion Irish Guards were at Aldershot. By this time the Brodrick Cap which was worn, as the Press put it, with the usual 'rake' over the right ear, had given way to the peaked version nearer the one in use today. Training was now carried out in khaki tunic, trousers and puttees, but drill was still performed in the white monkey-jacket, although this later fell into disuse when it was discovered that the constant inhaling of the pipe-clay dust caused silicosis and other chest complaints.

This was a heyday for the Irish Guards in the sports arena. Their team and individual boxers had an unrivalled reputation, and the champions of the three upper weights in the Service Championship were at one time all Irish Guardsmen. Their tug-of-war team was outstanding; and in their other main sport, athletics, the battalion were winning many laurels – and more a few years later when two young officers, Second Lieutenants the Hon H. R. L. G. Alexander (later Field-Marshal Earl Alexander of Tunis) and E. B. Greer (who was killed in 1917 commanding his battalion), became prominent in the track events.

The Irish Guards had their first taste of strike duty in the railway strike of 1911, which passed off peaceably enough. But trouble had been expected, and for the first time in their history the Brigade of Guards wore khaki in London; headgear was a problem, for the authorities reckoned that the soft forage cap would be useless against violent strikers – so it was decreed that they would wear bearskin caps. It would be hard to imagine anything more likely to provoke the baser instincts of a British striker.

In 1913 was adopted what is still the basic battalion organization in use today, for in that year the double-company system came into vogue and the previous eight companies were cut to four, each divided into four platoons and totalling some 100 men. At the same time the senior of the two Pay-Sergeants became known as the Company Sergeant-Major, and the junior, tht Company Quartermaster-Sergeant.

Over the years since their formation the Irish Guards had experienced that complete absence of religious or political controversy, or even debate, which has characterized the Irish regiments for so long. It was not that they did not deeply feel the unrest in their native country – it was just that they were above it; once in uniform they felt and owed an allegiance to a higher cause.

The years before the Great War were vital ones in the cause of Irish Home Rule. With Ulster armed – thanks to a brilliant coup whereby 35,000 rifles and three million rounds of ammunition were smuggled ashore at Larne and spirited away – and the South arming, the cause of amity, let alone peace, in Ireland seemed doomed.

The so-called Curragh Mutiny (which later came to be known more accurately as the Curragh Incident) – when a number of officers in the 3rd Cavalry Brigade under Brigadier-General Hubert Gough refused to march on Ulster, an order which they felt was imminent – had passed the Irish Guards by. It is true that two officers did march into the Commanding Officer to express their agreement with Gough's sentiments, but they were forcefully told to go away! Otherwise the Curragh Incident might never have happened for all the interest the Irish Guards showed.

On July 22nd, 1914, a meeting was held at Buckingham Palace with the Irish leaders to try to resolve some of the outstanding points of the Home Rule Bill. On its unsuccessful conclusion the two chief Irish protagonists, John Redmond and John Dillon, were walking past Wellington Barracks where were stationed the 1st Battalion Irish Guards. Some men spotted the two national heroes and soon the whole battalion was at the railings shouting, cheering and waving. Later the Commanding Officer, Lieutenant-Colonel the Hon G. H. Morris, spoke to the battalion about this remarkable and disgraceful demonstration, and he ended: 'You are all, or nearly all, racing men and like a good bet from time to time. Back what horse you like – but keep your tips to yourself.' They got the point.

But this incident was soon swallowed by one of greater importance to themselves and their whole generation. One week later Austria declared war on Serbia and, inexorably, the Great Powers were swept towards the abyss of the Great War.

Chapter 3

The Great War: Mons to the Aisne

On August 4th, 1914 – August Bank Holiday – Great Britain declared war on the German Empire.

The wildest rumours had been circulating early that day. Huge crowds roamed the streets and had gathered outside Buckingham Palace to cheer the King and Queen. There had been a feeling of exasperation with the government, but now, with the imminent prospect of war, that had been succeeded by an overwhelming sense of exhilaration. Inside the railings of Wellington Barracks the 1st Battalion Irish Guards watched the seething mass of excited humanity and went about their own business. But the unusual sight of Company Commanders with their Company Sergeant-Majors and Company Quartermaster-Sergeants in attendance doubling to the Orderly Room electrified them. Although they did not know it, the conference which then took place behind the closed doors was to decide the fate of all but a handful of those present in Wellington Barracks that hot August afternoon.

The door opened. The occupants emerged. It was war! The news spread rapidly and a great cheer went up from behind the barrack blocks. But there was little enough time for cheering as they were due to embark in eight days' time. The barrack gates were locked and then the Irish Guards got down to the immense and seemingly endless task of preparing themselves for the first campaign in their history.

Finally all was ready. On the morning of Monday, August 11th, the Battalion paraded under Colonel Morris, on the parade ground of Wellington Barracks, to be inspected by the Colonel of the Regiment, Field-Marshal Lord Roberts. It was the last time he was to see the Irish Guards, and the presence of their 'Bobs' that day, although he was looking very old and frail, set the seal on the glamour and excitement of their preparations for war.

The next morning, whilst most of London was still sleeping, they were addressed by the Major-General. 'Come back with basketfuls of medals,' he told them. Then, after a pack of playing cards had been distributed to every man, they marched from Wellington Barracks, before the envious eyes of the large number of Irish Guards Reservists who had to be left behind and who watched the departure of their 1st Battalion with bitter disappointment.

Over Vauxhall Bridge to Nine Elms Station they went, and thence by train to Southampton to join the growing flood of the British Expeditionary Force bound for France.

Thirty-two officers and close on eleven hundred rank and file – of which ninety-eight per cent were Irish and about ninety per cent Roman Catholics – embarked on the SS *Novara* that evening. She passed HMS *Formidable* off Ryde and exchanged signals with her. The battleship's last message to the Battalion was to wish them 'plenty of fighting'. And so the Irish Guards went to war.

They arrived at Le Havre early the following morning, to be met by an ineffective colonel, two Boy Scouts and a semi-idiotic interpreter. The business of unloading began. It had been a dreary, hot, crowded voyage, for the *Novara* was no pleasure ship, and an enforced wait in a hot, airless, tin-roofed shed while some missing men were rounded up improved no one's temper. They did not move off until close on midday.

On August 22nd, they marched into La Longueville and heard the sound of the guns for the first time. By a strange coincidence it was near this very spot that in rather more than four years' time those who were left were to hear the Cease-Fire.

It was a country of low hills and shallow valleys – a closely cultivated land of orchards and mile after mile of cornfields, the corn stooked ready for the harvest. The villages were clusters of white homesteads with red-tiled roofs, whose inhabitants and the children – especially the children – came out to welcome the men as they marched through. It was peaceful, and beautiful in the summer heatwave. The war still seemed very remote in those sweltering days of August, and the only thing out of the ordinary was the distant, ominous rumbling of the guns.

But the Irish Guards were more concerned with internal matters. 'The battalion marched well today', the diarist found time to record. And there still seemed to be endless time. News was sparse, and largely inaccurate. Perhaps it was as well that they did not know the full peril the British Expeditionary Force were in; but the trouble was that their commanders did not appear to, either.

On August 22nd the BEF started moving north towards Mons. That evening scouts from the 4th (Royal Irish) Dragoon Guards – the 'Mounted Micks' they were called – had a skirmish with some Uhlans (German cavalry) and carried the day. They, too, reported strong enemy forces moving down from the north. But this information, together with that supplied by some pilots of the Royal Flying Corps, who added that a further column believed to be a whole army corps strong was heading towards Grammont and thus threatening to outflank the BEF, was still disbelieved by GHQ. There were suspicions, too, that all might not be well with the French Fifth Army, but these were also ignored. Meanwhile, the BEF continued northwards into the noose which was quickly developing.

Even though the left flank of the BEF might be in the air, their right flank remained linked to the French Fifth Army. On the 23rd it became very apparent that this was no longer so, for the French were now in full retreat. Orders came for the leading elements of the British II Corps to take up positions along the Mons–Condé Canal. The advance had stopped; the retreat was about to begin.

As they trudged northwards through the heat and the dust, all this was unknown to the 1st Battalion Irish

Guards, who were part of 4th Guards Brigade of 2nd Division, one of the three divisions of Sir Douglas Haig's I Corps. On August 23rd the Brigade were ordered to take up a position to the south-east of Mons, and the 1st Irish, with the 2nd Grenadiers, marched forward, leaving the other two battalions of the Brigade, the 2nd and 3rd Coldstream, to dig in on some high ground a little to the rear.

Shaking out into artillery formation – small groups widely spaced apart – they advanced across the open fields. Bringing up the rear was the Regimental Sergeant-Major with two drummers holding the big bass drum between them, like a huge target. Though what possible use that could have been is not reported.

Shrapnel and rifle bullets buzzed over them. Innocuous, seemingly, and they were about to vote war a dull thing when one shell burst immediately above a cluster of men and two fell, wounded. Pressing on through a turnip field – where a covey of partridges exploded at their feet, frightening them far more than had the Germans so far – they found sanctuary in a ditch.

The Germans sounded very close now. It was as though a scythe of rifle and machine-gun bullets was passing directly over their heads. It came in gusts, whistling, snarling, sighing. 'This is Hell Fire,' said one; a good few were praying or crossing themselves. This was their first experience of the seemingly personal vindictiveness of enemy fire, and the apparent accuracy of the Germans astonished them: 'Jesus, and these are the fellows who they told us were militiamen.'

It was in this ditch that they 'dug-in' for the first time; the veterans later recalled these shallow 'trenches' with derision and dismissed them as examples of the 'valour of ignorance', as they put it.

They stayed there till late that night and then pulled back. But so inadequate was German Intelligence that the enemy never realized that they were gone until the following morning.

The position at Mons was now untenable. With the BEF

isolated on both flanks and opposed by two full German armies it was only a question of time before they were overrun. The retreat started.

With the Forest of Mormal in their path a split in the BEF was inevitable; so it came about that II Corps kept to the west of the River Sambre, while I Corps for the most part stayed on the east bank.

Taking part in the general withdrawal, the Irish Guards covered the initial retreat of 2nd Division and then pursued their own way south. They were joined around here by Drill-Sergeant Moran whose greeting 'F'what's all this talk of a retirement?' was received with laughter, and served to lighten the moment.

They bivouacked again south of the village of La Longueville forty-four hours after leaving the place – they had been without rest for the whole time. That evening the first mail from England arrived and was distributed by torchlight under the apple trees in the warm night. There was still a sense of unreality. No one knew what was happening. There seemed no reason for the retreat, and the explanation that they were conforming to the movements of the French Army was beginning to wear thin. Their true enemies appeared to be exhaustion and lack of sleep, and the dusty, irregular, badly cambered, pavé roads. But their plight was no worse than that of the other troops on that terrible march.

The following morning they continued their withdrawal, severely hampered by some troops of the French Fifth Army who happened to have chosen the same road. In late afternoon the column reached Landrecies, described as an unlovely, long-streeted town in close-cultivated country; to the north lay the forbidding mass of the Forest of Mormal, which had necessitated the dangerous separation of the two British corps. The battalion went into billets in the town along with the whole of I Corps Headquarters – a curious arrangement, and, in fact, there were no fewer than six general officers in Landrecies that night.

There had been rumours of the Germans being close on

their heels, but nothing had been seen of them and an alarm, sounded shortly after they had got in, proved false. Then at 7 PM they were attacked in earnest. The 3rd Coldstream were providing the pickets outside the town and, after the initial surprise, resisted the enemy resolutely. But

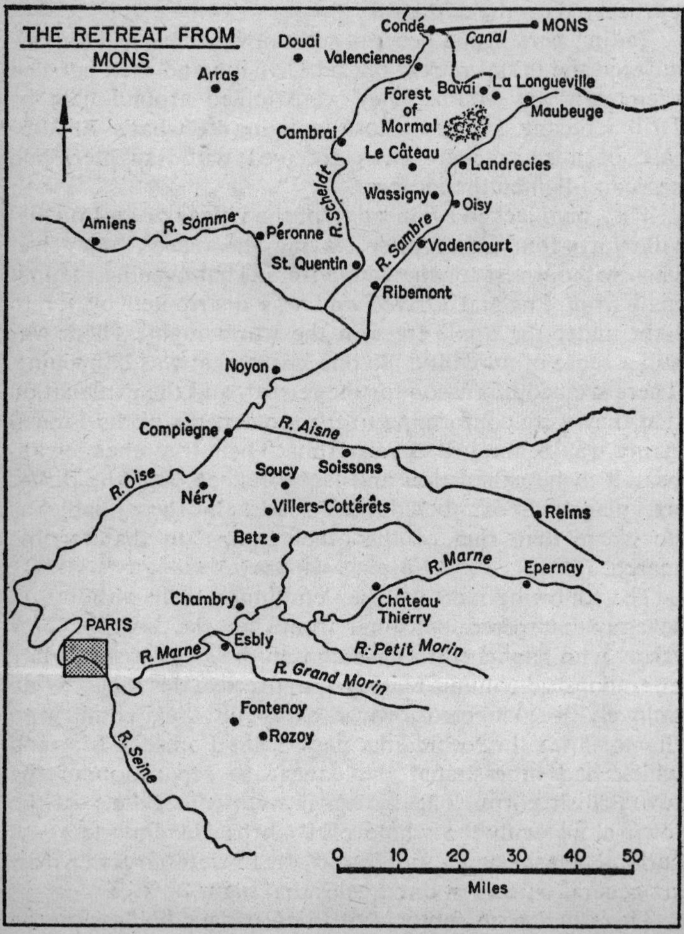

for a moment the situation seemed desperate, and in the town of Landrecies there was pandemonium.

While the rest of 4th Guards Brigade beat off the Germans, the Irish Guards were given the task of making barricades and loop-holing houses. Paving stones from the roads, miscellaneous carts – a few with their loads still aboard, to the fury of some officers who were looking for the stores they carried – tables, chairs, even a grand piano were added to the barrier at the end of the long main street of the town. The scene of unreality was helped by the blaze of a haystack which had somehow been set aflame, and the sight of a battalion cooker beating a precipitate retreat with its fire still alight and leaving a trail of sparks behind it. At some stage it was recorded, too, that the bass drum was run over by another cooker and seen and heard going down the main streets of Landrecies strung on the foreleg of a horse. Certainly the drum was never seen again after that night.

The Corps Commander and most of his staff left in a hurry. 4th Guards Brigade withdrew more leisurely the following morning. The engagement, variously reported as being anything from a brigade downwards, was, in truth, little more than a probing attack, for the German troops had no idea that Landrecies was occupied and had been told to bivouac there themselves for the night.

It was shortly after they left Landrecies that 4th Guards Brigade passed another brigade of their division fallen out and resting by the side of the road. Instinctively, and with no command given, the men came to attention, mouth-organs were pulled from pockets and to the tunes of their respective regimental marches the exhausted and dust-covered men marched past their comrades exactly as if they were on Guard Mounting. Such was the power of discipline and the ingrained pride of regiment. It was an intensely moving and uplifting experience for those who witnessed it.

So far II Corps had taken the weight of the German effort. Separated from I Corps by a gap once as wide as eighteen miles, they had the enemy to contend with as well as the heat and the dust. Action after incredibly gallant action fought off the Germans snapping at their heels. The enemy were soon made to respect the fighting powers of the BEF, but the continual marching, lack of news and play of fear on the imagination of tired men inevitably started to tell even on what is believed to have been the finest army ever to leave these shores.

There was no panic, but reactions were slower now, and men trudged on, holding their ranks almost instinctively and as a form of self-protection, for to break them was to lose the cohesive effort of keeping on. But the worst was now nearly over, although they were not to know it, for the enemy had over-stretched themselves.

Nevertheless, the Germans were still following up, believing the BEF to be in complete disorder, and on September 1st in the wooded country south of Soissons, around the town of Villers-Cottérêts, they caught up with the rear-guard of I Corps, then being provided by 4th Guards Brigade. (See map, p 24.)

The Irish Guards breakfasted on September 1st amongst wet lucerne and dripping corn, for there had been a very heavy dew that night and a dismal, thin rain was falling. Behind them stretched the dark, damp beech forests of Villers-Cottérêts. They took up a position outside the wood. One of the companies was in a graveyard, and there they sited themselves amongst the tombstones and loop-holed the walls of the enclosure. To their left were the 2nd Coldstream, while in the depths of the forest the 2nd Grenadiers and 3rd Coldstream had taken up another line of defence along a ride.

Around nine o'clock some Uhlans were spotted trying to infiltrate into the forest behind them. Some well-directed fire made them hastily change their minds. A brief lull ensued and Colonel Morris, who was in command of both forward battalions, thinking that they would be unmolested for a bit, ordered the 2nd Coldstream to pull back from their very exposed positions. He was about to do the same with his own battalion when a message came from the Brigade Commander with orders to hold on. The Coldstream were by now out of touch, so the Irish Guards prepared to receive the next attack, but alone.

The enemy closed fast, and had even managed to bring up a gun which fired shrapnel low over their heads. The Irish Guards held them out of the woods for a spell and the two Maxim guns did great execution, but soon the enemy artillery and machine-gun fire gained the upper hand. The Irish were forced to withdraw to get what protection there was in the forest itself. At first the disciplined shooting gave the Germans cause to hesitate, but soon they found the gaps between companies and platoons and realized that both flanks were in the air. Then the situation became tricky.

Fortunately the Germans had elected to sweep through the woods rather than keep to the rides. Thus they, too, lost all cohesion. The woods rang with the sound of little hunting horns and cries of 'Hoch, hoch'; clearly the enemy were having their troubles too.

Throughout this time the Commanding Officer was riding up and down smoking cigarette after cigarette in a long holder, chaffing and giving encouragement to his men, until, as was inevitable, he was killed. In the woods the air was stifling; the heavy undergrowth hid wounded and advancing Germans impartially. Control became well-nigh impossible, so when the order to withdraw was passed, there were some who did not receive it. The remainder started a steady retirement, every now and then turning on their pursuers to keep them at a respectful distance.

The Grenadiers and the two Coldstream battalions holding the reserve line were attacked from both sides at about the same time. They resisted stubbornly, while the Irish worked their way steadily back towards them. Then all four battalions, intermingled as they were, proceeded to beat off the enemy throughout that long, hot day. They wore different cap-stars but they fought not so much as a mixture but as an amalgam; their common training and common discipline stood them in good stead, and officers took command of any men they could find in the gloom until, eventually, the enemy had had enough and 4th Guards Brigade could make their way back and out of the oppressive forest of Villers-Cottérêts. The Grenadiers lost four officers and 160 men; the two Coldstream battalions three officers and 37 men; the Irish Guards nine officers and 115 men, including the Commanding Officer, Second-in-Command and one Company Commander killed and their Adjutant wounded. The Irish Guards had been blooded, but it was a bitter price to pay.

Their greatest loss was their Commanding Officer, who knew the temper of his men so well. Colonel Morris had been a hard man, but a just one; a martinet with a strong Irish accent. Before the war he had instituted a weekly meeting when the men could make any complaints or suggestions they might have on the standard of food or other welfare matters. This was a revolutionary step in those days. The other regiments of the Brigade copied his example, and in time the rest of the Army followed suit. He had been a fine Commanding Officer.

They reached Betz at eleven o'clock on the night of September 1st to be greeted by the sight of a burning ration dump set alight by the orders of some over-anxious staff officer, and for once received no supplies. Nor the next day, when they arrived at Esbly on the River Grand Morin where they bathed and were able to take their boots off their swollen feet – and where a good few couldn't get them back on again. They had marched 220 miles in seventeen days.

The retreat was officially over. The Germans had shot their bolt, and with the linking of I and II Corps they had anyway missed their chance. For they, too, were exhausted. The armies paused.

But not for long; for on September 6th the BEF turned on their pursuers in an advance that was to take them sixty miles to the River Aisne before the Germans managed to recover their poise and bring them to a halt. This was to be the end of the fluid war on the Western Front; thereafter success was measured in yards, and triumph was an enemy trench taken and held.

The early days of September saw the Irish Guards advancing north-east. 'All along the way were wounded men, abandoned guns, broken ammunition wagons and general helter-skelter,' wrote one private. In Rozoy the church bells were ringing and the priest was standing in his doorway. As the battalion passed he gave them his blessing.

Their contribution to the Battle of the Marne was to help force a crossing of a smaller waterway, the Petit Morin, to the south of the main river. There was well-organized resistance, but their main worry were the snipers who, many perched in trees, were able to pick off the officers, distinctive in their gold-peaked service caps. In this affair of the Boitron Woods, as it came to be known, the Coldstream captured three officers and some ninety men with six machine-guns. Not wishing to be encumbered they passed them on to the Irish Guards in reserve who promptly paroled the officers and invited them to dinner. The niceties of war were still being observed.

The first batch of reinforcements arrived on September 18th, and the battalion reorganized. They were beginning to realize that 'home before Christmas' was a myth. The next day Lord Cavan came as Brigade Commander to take over from Brigadier-General Scott-Kerr, who had been wounded in the Villers-Cottérêts battle.

Choice of a new Commanding Officer was proving something of a problem. On hearing of the death of Colonel Morris, Brigadier-General C. FitzClarence, VC, then commanding a brigade in Ireland, offered to revert to lieutenant-colonel in order to command his regiment, but he was expressly forbidden to do so – he was required for the 1st Guards Brigade as it later transpired, and was killed at the First Battle of Ypres. So Lieutenant-Colonel Lord Ardee, of the Grenadiers, was appointed. With him came Captain Lord Francis Scott, also of the Grenadiers, and Captain N. Orr Ewing, of the Scots Guards. These were the first of a long line of senior officers loaned by the generosity of the brother regiments of the Brigade to make up the depleted strength of the Irish Guards.

Chapter 4

The Great War: First Ypres to Festubert

The Aisne advance ground to a halt. Focus was now on the Flanders front, 120 miles to the north-west. As the Germans tried to envelop the left of the line so reinforcements were rushed in and the gap was sealed. The Allied armies now extended from the Channel to the Swiss frontier, and the scene was set for the attrition of trench warfare. But some reorganization was necessary, for the BEF was sandwiched between the Fifth and Sixth French Armies, and further to the left lay another British force. It was decided to concentrate the BEF on the Flanders front. So, at the beginning of October, the big switch commenced.

On October 9th Antwerp fell to the Germans and the troops released joined those already opposing the BEF. The original intention had been to mount a major offensive on this front, but, as more and more German reserves poured in, the BEF was very soon fighting a desperate defensive battle.

By the time I Corps reached the scene the situation was critical. No organized take-over of a sector was feasible; as units arrived they were flung straight into the fray, and there they had to fight against heavy odds. The First Battle of Ypres, as it came to be called, was perhaps the most crucial engagement of the whole war. As Kipling put it: 'Companies had to do the work of battalions, and battalions of brigades, and whose only relief was a change of torn and bloodsoaked ground from one threatened sector of the line to the next.' Only the incredible staunchness of the British troops staved off disaster utter and complete. But they paid a terrible price for their victory and lost the cream of the pre-war Regular Army – a handicap which had its repercussions at all levels until near the end of the war, by which time a new generation of officers and men had been trained and blooded, and the veteran was an oddity.

On October 21st the 1st Battalion Irish Guards marched through Ypres and were allowed a few hours' respite. At that time Ypres was undamaged and the townspeople came out to welcome the troops and give them coffee and hot potatoes. In return the Irishmen amused their hosts by dancing jigs on the pavement. In the midst of the gaiety a squadron of French cavalry, resplendent in plume and cuirasse, came through at the trot, to add a colourful touch to a scene that was soon to become a sad and nostalgic memory for those few who were ever to see 'Wipers' again. The news of their religion soon carried, and as they left Ypres the nuns gave out rosaries as they passed.

The three divisions of I Corps occupied a salient from Zonnebeke in the north, via Reutel across the Ypres–Menin road to Zandvoorde, with 7th Division on the right, 2nd Division on the left and 1st Division in the centre. The junction of the 1st and 7th Divisions was on the Menin road, and it was from this point that the main enemy thrust came.

On October 25th the whole of 2nd Division was ordered to counter-attack in the direction of Reutel in order to relieve the pressure which was building up against their right-hand neighbours. The attack of the 1st Battalion was through the thick undergrowth of Reutel Wood, known as Polygon Wood, which held them up badly. They all but reached the village of Reutel itself but were finally pinned down some three hundred yards short, tantalizingly close to the houses and the church. They had been ordered to hold ground gained at all costs, but as darkness fell their position was perilous indeed, for just before this they had lost touch with, and were also well forward of, the nearest friendly unit. The Germans suspected that they were on their own, well ahead of any support, and probed unceasingly in the heavy rain, but the battalion managed to link up with the Worcesters some six hundred yards to the rear by means of a double line of men almost back to back. The little salient held, but there was no sleep for anyone that night.

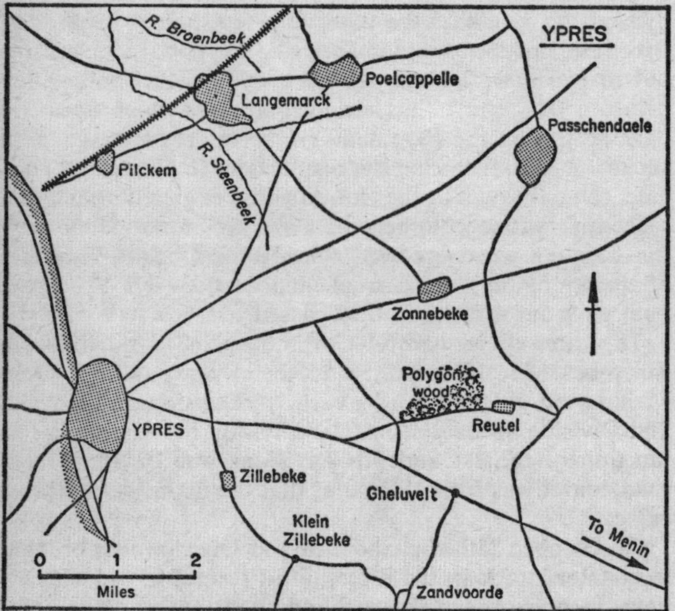

At dawn the 2nd Coldstream managed to force their way through the thick woods to their help, and together the two battalions tried again to take the offending village of Reutel. A few farmhouses on the fringe fell, but after heavy losses from enfilading machine-guns even these could not be held. The diary records that the night was memorable in so much as the men, who had not eaten for forty-eight hours, were allowed to consume their emergency rations. They were relieved the following morning.

The next days were punctuated by a continuous series of alarms, unsatisfactory to everybody, expensive in blood and exhausting to nerves and everything but spirit. A big attack had been expected on the 29th, attended by the Kaiser in person, or so they were told, but the battalion had a peaceful day. Their peace was short-lived.

Crisis day of the First Battle of Ypres was October 30th. Called up in the morning to dig some reserve trenches in the area of Polygon Wood, in the afternoon the Irish Guards and 2nd Grenadiers were ordered to stop digging immediately and go to the help of the sorely pressed 7th Division near Klein Zillebeke. There they found a gap some quarter of a mile wide dominated by a small wood to the east. They dug in all night, helped by the light of a steadily burning farmhouse which silhouetted the enemy, conspicuous in their spiked helmets.

At dawn they were ready – but hardly for the barrage which broke about them. Black Marias and Jack Johnsons – both heavy shells – shrapnel, the whizz-bangs from the German 4.2s – they got the lot. They were shelled continuously from seven o'clock in the morning to eleven o'clock that night. As the day wore on so the intensity of rifle and machine-gun fire also grew to horrendous proportions. In the beginning they had been supported by a battery of the Royal Field Artillery, but when on the left the flanking battalion was forced to fall back, even this meagre assistance 'limbered up and went off as though it were the Military Tournament' as one officer described it. Casualties in the totally inadequate trenches were very heavy.

There was no let-up in the shelling the next day either. One trench, holding a full platoon of Number 3 Company, received a direct hit – there were three survivors. In the distance a band was playing *Die Wacht am Rhein* – we'll give them *Watch by the Liffey*, said someone; and then came the attack. With support from some close-range artillery weapons, and to the accompaniment of a crescendo of small-arms fire, the Germans hurled back the remnants of Number 3 Company. Numbers 4 and 1 Companies on the right and left respectively fell back on the reserve company occupying rough trenches to the rear. There they made their stand, helped by one platoon which had been unable to pull back from the original trenches and which took the full weight of the German onslaught by holding off the enemy from the main battalion position precariously situated on the edge of Zillebeke Wood.

Every man able to bear a weapon was called in; cooks, orderlies, the lightly wounded, all took rifles, and with their backs to the wood fought off the Germans. 'The Irish Guards remained to suffer further losses from the enemy's continuous artillery bombardment...' So read the official account. One who was there put it another way: 'It was like a football scrum. Everyone was somebody, you'll understand. If he dropped there was no one to take his place. Great days! And we not so frightened as when it came to the fighting by machinery on the Somme afterwards.'

The situation was so critical that the Adjutant was sent off to Brigade Headquarters to get help. It arrived – eventually. The thin line was still holding out but had suffered terrible casualties, including the Commanding Officer, Lord Ardee, wounded. The battalion had eaten no food for thirty-six hours.

On November 2nd they reorganized into three under-strength companies. On the 7th they lost another Commanding Officer, Major H. A. Herbert-Stepney. On the 8th they reformed into two small companies. On the 9th they could muster only four platoons. On the 18th, after further losses, they were pulled back for the last time. They were relieved by a company of the 3rd Coldstream – and only one company was needed.

The few survivors barely recall the next few days. The relief from the all-pervading shelling was complete anti-climax. It took two days to get back to the promised haven; they arrived eventually, eight officers and 390 men in all – they had entered the battle over a thousand strong – desperately tired, ragged but cheery, in a cold that froze the water in their water-bottles. The diary, with warranted pride, added 'not a man fell out'.

The whole of 4th Guards Brigade had suffered staggering casualties. The Irish Guards' share alone was close on a thousand all ranks killed, wounded or missing. The battalion had been practically wiped out, partially rebuilt and nearly wiped out again. The remnants were exhausted in mind and body; only their morale and irrepressible humour kept them going. The refit when it came was complete from trousers and groundsheets to mittens and mufflers generously provided by His Imperial Highness the Grand Duke Michael of Russia.

But their sacrifice had not gone unnoticed. For their Brigadier, Lord Cavan, wrote to the Commanding Officer: 'I want you to convey to every man in your Battalion that I consider that the safety of the right flank of the British sector depended entirely upon their staunchness after the disastrous day, November 1st. Those of them that were left made history, and I can never thank them enough for the way in which they recovered themselves and showed the enemy that the Irish Guards must be reckoned with however hard hit.'

During the pause they were joined by the Reverend Father Gwynne, the first Roman Catholic Chaplain appointed to the Regiment and the forerunner of a long line of wartime chaplains who hold a unique place in the love and life of the Irish Guards.

The call for the priest was frequent in the days that followed. Pressing forward as far as he could get, Father Gwynne was always to be seen where the excitement was at its most intense. So much so that the Commanding Officer had to give him a direct order to keep back, but, as they put it, it was amazing how often he happened to forget. On one occasion, exhausted in mind and body, he fell into a deep sleep in one of the forward trenches which was being prepared under fire. He was roused by a respectful whisper from one of the working party: 'We've dug to your head and your feet, Father, and now, if you'll get up, we'll dig out under the length of you.' And it was Father Gwynne 'the brave little priest' who died in 1915 and was last seen kneeling by the side of a dying German.

To the men the presence of the priest was a wonderful and reassuring one. To the Commanding Officer he was a tremendous source of strength and an invaluable sounding-board to the currents which flow beneath the surface of any organized body of men.

Before they returned to the line of November 14th they received word of the death of their Colonel, Lord Roberts, who had contracted pneumonia during a visit to the front. Field-Marshal Earl Kitchener was appointed in his place as Colonel of the Regiment.

Trench warfare in its true version was now becoming the accepted pattern. Constant digging and 'improvements' became the normal occupation. Wire was scarce, so they hung bully-beef tins on string and threw out others on the ground in front so that any advancing German would kick them and tell of his presence.

Sandbags had by now made their appearance, one for each man. 'To be carried on the soldier,' read the order authorizing their issue, 'and in very cold weather permission is given for these to be used as foot muffs.' Other refinements of war were appearing as a result of the new-style fighting. Close-quarter combat and 'bombing' were new and rudimentary arts as yet. The early 'bombs' were made out of jam-pots as lethal to the thrower as to the intended victim.

The aeroplane had already made its presence felt and 'aeroplane duty' was a drudgery employing a full company at a time whose task it was to keep on the look-out for enemy aircraft. They were then allowed to open fire with every weapon that could bear. Our own pilots had troubles enough dealing with the technical unreliability of their aeroplanes as well as the enemy. They found it disconcerting, and said so, to be met by fire from their own side as well.

The Christmas truce of 1914 which received so much

publicity in the British Press hardly affected them at all. Life proceeded as normal with the occasional heavy bombardment adding unwelcome spice. Christmas cards from the King and Queen were distributed, and pipes and tobacco with her Christmas cards from Princess Mary.

Life out of the line assumed a pattern which was to become stereotype; sleep, clean and drill was the sequence, then came route marches, rifle exercises, more drill, kit refits and inspections, the learning and practising of new techniques, and the ever-recurring task of absorbing new drafts which came in a steady stream from the holding battalion at Warley. By now all sorts of luxuries were reaching the front as people at home became geared to war. At the end of January the battalion returned to the line at a place described unofficially as 'not very wet but otherwise damnable'. This was among the brick-fields near La Bassée, south of Neuve Chapelle.

The country was flat, without feature or attraction. Their line was centred on a collection of a dozen or so stacks of bricks some thirty feet high, some held by us, but the majority as securely in the hands of the enemy. The area was interlaced with underground workings, saps and communication trenches. 'The whole position is most interesting and requires a considerable amount of ingenuity,' pronounced one officer; the comments of the majority are not recorded.

Bombing attacks and local assaults became the order of the day. Cheek-by-jowl living with the Germans – in places less than ten yards away – became trying to the nerves. And no sooner were the battalion installed in their solid Flemish billets than the pattern of their lives for the next weeks emerged with frightening suddenness.

For early on the morning of February 1st, 1915, the enemy rushed and took the forward trenches held by the 2nd Coldstream. That unfortunate battalion were stretched to the limit and already holding a huge sector, so Number 4 Company of the Irish Guards were ordered to assist in the counter-attack. The Germans were adept at making good their captures, and the counter-attack gained a foothold but no more. With all the officers of the company either killed or wounded, command devolved upon CQMS Carton, who, in spite of verbal orders to retire 'which he did not believe', held grimly on until morning with the survivors of his company and any stray men who came along.

The counter-attack was continued at dawn by Number 1 Company and the remnants of the Coldstream. The 'heavies' were brought in, and as their barrage concluded on and behind the enemy lines it was succeeded by a vicious ten minutes' 'rapid' from the rest of the battalion and anyone else with weapons to bear. The parapet of the opposing trenches was simply blown away as the riflemen laid down a tremendous fire. When the magazines were empty, loaded rifles were passed to the firers and they continued. It was calculated that each of 140 riflemen fired off 200 rounds in that astonishing ten minutes – and such figures were far from unusual in the BEF in those days; no wonder the Germans thought they were faced by machine-guns. In the middle of it all, and in a gesture of utter despair, a German officer mounted on a horse rode from one of the trenches in a vain attempt to rally his men. He and his horse disintegrated in seconds. And then the assault went in.

It was here that Lance-Corporal O'Leary won his Victoria Cross. 'He rushed up along a railway embankment above the trenches,' reads an unofficial report, 'shot down five Germans behind their first barricade in the trench then three more trying to work a machine-gun at the next barricade fifty yards farther along the trench, and took a couple of prisoners. Eye-witnesses report that he did his work quite leisurely and wandered out into the open, visible for any distance around, intent upon killing another German to whom he had taken a dislike.' By his great gallantry he undoubtedly saved the two leading companies of the Coldstream and Irish Guards a great many casualties.

The attack was successful, the position retaken. It had cost them five officers and thirty-four men killed, wounded or missing. An isolated, costly, yet all too typical incident in the close-quarter world of the brick-stacks.

When his award was announced O'Leary became a national hero. Promoted to sergeant before the whole battalion, he returned to his investiture and then to Ireland to be mobbed. A slight, self-effacing man with tremendous charm, he became the darling of All-Ireland. The first Irish VC in the war, and the first for the Irish Guards since the formation of the Regiment, he featured in a remarkable recruiting poster (see illustration section). He and his mother and father conducted a private recruiting campaign, the like of which has never been seen before, or since. An O'Leary Fund was set up to look after the widows and orphans of Irish troops; July 11th, 1915, was set by as O'Leary Day, when a massive recruiting drive was launched with the great man as its figurehead; and for a time the Irish Guards were known as 'O'Leary's Lads'. Throughout it all, O'Leary maintained his modesty and happy-go-lucky view of life. Commissioned into the Connaught Rangers on his return to France a few months after his great deed, he finished the war as a captain. Major Michael O'Leary, VC, died in 1961.

By the time the Prince of Wales came to visit the front in late February, they had grown accustomed to the peculiar life of the brick-stacks. The scares of enemy mining parties were trying to the nerves and induced the formation of a mining company of their own, but apart from one abortive effort by the Germans to retake those stacks they had recently lost, the month passed as peacefully as frost and the enemy would allow.

They moved into Corps reserve at Bethune where the drudgery of trenchwork was succeeded by 'steady drill and route marches' – a football competition for a cup presented by the Bishop of Khartoum – which they lost – and that wonderful experience of 'first leave'. Here, too, they celebrated St Patrick's Day.

The initial reinforcement draft composed exclusively of former members of the Royal Irish Constabulary arrived shortly afterwards. The government had given permission the previous November for a limited number of Constables to be enlisted in the Irish Guards. Their quota was only two hundred, but nine times that number volunteered. By the end of the war nearly seven hundred former RIC men had served with the Irish Guards – of these some ten per cent, having passed through the ranks, gained commissions in other regiments. Splendid material, the first detachment were all well over six feet tall, magnificent men who soon made their mark. And with their welcome addition the battalion could now muster twenty-six officers and over a thousand men.

They heard with horror of the German gas attack at the Second Battle of Ypres, but were not called upon. Their turn came quickly enough, though, at Festubert.

Intended to discourage the Germans from reinforcing their armies opposing a massive French attack towards Lens, away to the south-east, Festubert was the first experience the Irish Guards were to suffer of the true futility of the type of war they were fighting. Like those battalions who had gone before them over the same blood-soaked ground, they were caught by enfilading machine-guns on what had once been flat and open fields, and cut to ribbons. Of the twenty-eight officers they had at the beginning of that day, only two were untouched; of the 1,100-odd men they mustered at the start, nearly 400 were casualties. As Kipling wrote: 'They came through it all, defeated, held down at long range, but equable in temper and morale.'

It was fortunate the flood of manpower was at its height in the summer of 1915 to make good their losses. Even when the seemingly endless appetite for reinforcements for the 1st Battalion had been satisfied there were men and enough to spare.

Since those far-off days when the 1st Battalion had marched out of Wellington Barracks, the 2nd (Reserve) Battalion had been fully employed – welcoming, equipping, training, and dispatching the reinforcements for the 1st Battalion.

The first few days after the departure of the 1st Battalion had been the worst. Then, when the Reservists were still flooding in and with only a tiny skeleton staff to deal with them – for all but a few officers had left for France – the men, embittered at being left out of the fight, became difficult to handle.

By the evening of the first day, by which time those left behind had moved to the Duke of York's Headquarters in Chelsea, the situation was getting nasty. The gates had been shut sometime before in an effort to keep down numbers inside to manageable proportions, but this merely antagonized those outside and did nothing to cool the tempers of those within. Then someone got hold of some alcohol and crisis point was reached.

At length the Major-General, Major-General Sir Francis Lloyd, received word of what was afoot. He arrived in a towering rage and addressed the men, lashing them with his tongue in a way they had never heard before. He told them they were a disgrace to the Guards; that he never expected to see a battalion of Guards behave in this way; that he regretted the day the Irish Guards had been born. They listened to him spellbound and fascinated. At length, spent with his own vehemence, he paused. There was silence for a few moments, then a voice at the back shouted: 'Three cheers for the General' – they roared. He later wrote: 'What can you do with such men; how could I but love them.' There was no more trouble, and only too soon those who had thought themselves left out of the war were in the thick of it.

With the loan of a number of officers and NCOs from the other Regiments of the Brigade, who were suffering no such shortage of key members, the 2nd (Reserve) Battalion came into its own and moved to the barracks at Warley; and really old barracks they were. One time used by the Honourable East India Company and first condemned fifty years before, they were hardly fit to live in. On his first round of inspection the Commanding Officer, detecting a peculiar smell and suspecting the drains, sent off the Quartermaster to investigate. The latter returned. 'Couldn't be the drains, sir,' he said. 'There aren't any.'

It was at Warley that, on July 15th, 1915, they heard that they were to become a service Battalion, the 2nd Battalion Irish Guards, and to join the then forming Guards Division. No longer were they to be concerned with the endless training of men for an impersonal sacrifice; instead, they were to be now a breathing entity themselves. They had the example of their 1st Battalion before them, but they had no traditions themselves; these they made up as they went along. Although the 1st Battalion could outmatch them in experience they were their equal in spirit and soon they were to show that they were their equal in battle.

Under the command of Lieutenant-Colonel the Hon L. J. P. Butler, the 2nd Battalion sailed for France on August 16th. They took with them their newly formed Pipe Band, whose first set of pipes had been presented shortly before by John Redmond. On August 30th the historic meeting of the two battalions took place at St Pierre, south of St Omer. They ate their dinners together, when the veterans took pains to point out how shiny were their cap-stars and advised them not to betray their rank under fire, adding darkly that they would soon know the reason why. Captain Charles Moore, a noted horse expert – and until his death a few years ago, Her Majesty the Queen's racing manager – cast a few disparaging remarks about the 1st Battalion's horses and then, under the critical gaze of the veterans, the new battalion moved off. It was a nervous moment. For their part the 1st Battalion could find no fault, although they immediately christened their comrades 'The Irish Landsturm'. And so the two battalions parted; they were to meet again soon under less happy circumstances.

Chapter 5

The Great War: Loos to Ginchy

The newly formed Guards Division assumed its final shape in the summer of 1915. The wisdom of creating such a large élite formation was discussed acrimoniously and at length. There had been a Guards Division a hundred years before at Waterloo, but in the South African War the brigades formed by the six battalions of Guards had been kept apart so as to act as leaven to as many formations as possible. That is how many pundits felt the Guards should be handled in this war. They were overruled.

A fourth battalion of Grenadiers was raised; the Irish had been increased to two battalions, and with the advent of the 1st Welsh Guards, raised on St David's Day, March 1st, 1915, a total of twelve battalions could be mustered. These were organized into three full brigades each of four battalions and for this purpose the two Guards Brigades, 1st and 4th, both already in France, were incorporated. The Household Cavalry provided the divisional cavalry – still deemed necessary – and the cyclist company. And a thirteenth battalion was included; this was the 4th Battalion Coldstream Guards (Pioneers) whose very necessary task it was to construct field works and keep in repair the roads behind the divisional front.

The first Commander of the Guards Division was Major-General the Earl of Cavan, a Grenadier. Lord Cavan had commanded 4th Guards Brigade with great distinction through the thick of the winter and spring fighting. Then had been sown a seed of mutual affection with the Irish

Guards – whose Colonel he was later to become. An outstanding and highly regarded commander, he was trusted by all. There could have been no more popular choice.

For 4th Guards Brigade the departure from 2nd Division with whom they had served for nearly twelve months was a sad affair. Between lines of cheering men they marched past their old divisional commander, Major-General H. S. Horne, for the last time. Perhaps the saddest farewell of all was the parting from the 1st Hertfordshire Territorials –

ORGANIZATION OF THE GUARDS DIVISION (1915)

1st Guards Brigade	*2nd Guards Brigade*	*3rd Guards Brigade*	*Divisional Troops*
2nd Battalion Grenadier Guards	3rd Battalion Grenadier Guards	1st Battalion Grenadier Guards	4th Guards Pioneer Battalion, later 4th Battalion Coldstream Guards (Pioneers)
2nd Battalion Coldstream Guards	1st Battalion Coldstream Guards	4th Battalion Grenadier Guards	
3rd Battalion Coldstream Guards	1st Battalion Scots Guards	2nd Battalion Scots Guards	*Cyclist Company* found by the Household Cavalry
1st Battalion Irish Guards	2nd Battalion Irish Guards	1st Battalion Welsh Guards	*Divisional Cavalry Squadron* found by the Household Cavalry

their 'Herts Guards' as they called them. This Territorial battalion had joined the Brigade in the previous December and had fought alongside them ever since, carving an imperishable name in the process. They were affectionately known to the Micks as our 'Second Battalion', and by the Coldstream with no less affection as their 'Reserve Battalion'. There were no closer friends. (In February 1970 the surviving members of the 1st Herts Territorials presented the lock of the gun which Lance-Corporal O'Leary, VC, had captured, to Regimental Headquarters.)

The new division assembled at St Omer. By the end of August they were complete and their training started in earnest. They tried the new version of war in the next weeks. They dug trenches by day and practised relief procedures in them by night; they 'attacked' villages; they practised 'bombing' until they were perfect at it; they tried out their new Lewis machine-guns; they rehearsed the role they were to play in what was declared would be the 'biggest battle in the world's history'.

By mid-September all was ready; field-training gave way to ceremonial drill and Lord Cavan inspected his brigades on the 16th and 17th of that month.

The planned advance was to be on the largest scale yet: to the south, General de Castelnau's group of French armies in the Champagne country east of Rheims were to attack in concert with French Tenth Army, and the I and IV British Corps in the area around Lens. It was indeed a tough task, for there had been plenty of time for the Germans to prepare themselves. Their two lines of defence were immensely strong; sited with great care and skill, with a sophisticated system of support and communication trenches. The Irishman who remarked that 'Jerry must have built them with the idea of staying forever' was nearer the mark than he imagined. The specially constructed shell-proof redoubts which could sweep the flat ground with enfilade fire, the triple lines of barbed wire and the wealth of reserves which the Germans could pour in at a moment's notice, rendered the whole operation a formidable undertaking.

Gas was to be used by the British for the first time; and a four-day preliminary bombardment by a thousand guns – more than had ever been used before but in the event nothing like enough – was thought would do the trick. Yet the Battle of Loos, the first major British attempt to end the stalemate of trench warfare, fizzled, failed and ended in disappointment after a few thousand yards of carnage-covered ground had been won.

To the south-east of Loos itself the gradual slope of Hill 70 dominated the surrounding plain. It was here and in the Chalk Pit area to the north that the Guards Division were committed for the first time. Their attack was ordered for the afternoon of September 27th. The objective of the 2nd Battalion Irish Guards was Chalk Pit Wood, a collection of scruffy trees which terminated in a deep quarry. Under the gaze of some 1st Battalion officers, who had clambered on to the wreck of a water-tower, two companies advanced and, thanks to heavy smoke, managed to secure their objective without much trouble. (See map, p 47.)

As the smoke cleared the Germans retaliated. There was little cover, and the combination of shrapnel and withering machine-gun fire from Hill 70 to the south, which dominated the new positions, halted the attacking companies in their tracks. The 1st Scots Guards, with a few Irish who had somehow attached themselves, gallantly fought forward and managed to reach a mine working, Puits Number 14B, just south of the Chalk Pit itself, but being unsupported had to fall back. The Irish Guards got into and beyond Chalk Pit Wood, but finding themselves bereft of any cover on the far side, in full view of the enemy and without artillery support, they, too, had to fall back. 'Jerry did himself well at Loos upon us innocents. We went into it knowing no more than our own dead what was coming, and Jerry fair lifted us out of it with machine-guns. That was all there was to it *that* day.'

The watchers on the water-tower had seen it all: seen their 2nd Battalion advance 'keeping their direction and formation perfectly' into the wood, had sensed the momentary elation, and then seen them fall back behind it. All too few of them this time. The 2nd Battalion Irish Guards lost eight officers and over 300 men in this their first battle. And casualties in the other battalions of the division were as heavy. But by their conduct that day the Guards Division earned the admiration of the Army and laid the foundations of their tremendous reputation. From then on the measure of perfection in whatever required doing was the 'Guards' way'.

The fighting dragged on into 1916 and a period of bitter, tedious trench warfare followed. A period of reliefs at night and incessant shelling by day, of mud and exposure, of gangrene, exhaustion and death. Unpredictable death from a stray shot, or, as it was calculated, inevitable death in a planned attack. It all in some way appealed to the morbid fatalism of the Irishman, but he never let it undermine his spirit. But static warfare was against his nature. 'There's no spark in it like the rough and tumble days of Landrecies,' was how one put it.

More trying was the constant work. The tasks within the lines, wiring, improving dugouts or fire-trenches, sand-

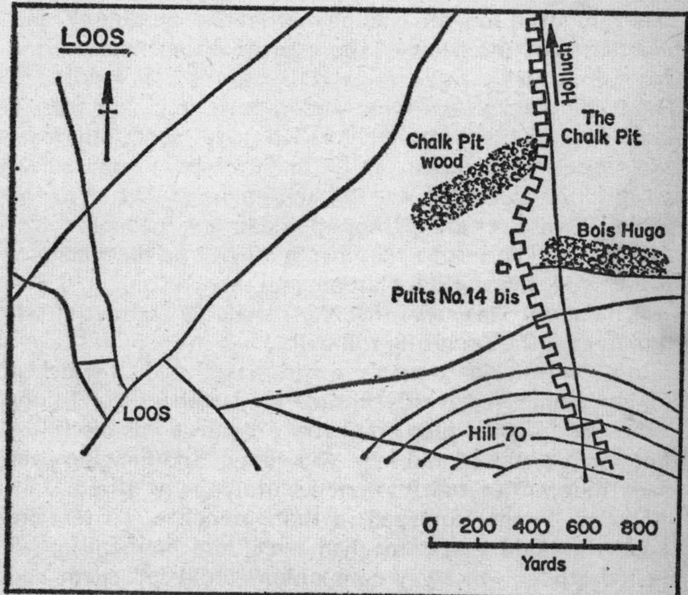

bagging, reinforcing the parapets, constructing fire-steps or carrying ammunition were all of immediate personal importance; but the constant preparations in the rear areas, repairing roads, unloading transport or burying anonymous dead, were the ones which sapped the spirit and mortified the soul – 'men take more notice, you'll understand, of one extra fatigue, than any three fights' summed it all up. County council work they called it, and it was slight consolation to know that the trenches they were so laboriously digging in the rear areas constituted the seventeenth line of defence! But it was a matter of pride that any trenches they occupied had to be as immaculate as hard work could make them. It was commonly known that anyone could tell that the Guards had been occupying a trench by its orderly appearance, and the cynics in the division reckoned that they were only sent to an area if the trenches needed remaking.

Within the lines life assumed a fearful monotony, but punctuated by moments of sheer terror. From time to time they went over the top – 'popped the parapet', they called it. The technique was the same; before zero hour – or what a German once inappropriately called 'love' hour – the barrage started. Designed to cut the enemy wire on some selected piece of ground, it was of such intensity that 'even the worms themselves are getting up and crying for mercy', declared one. Then came the charge when one machine-gun against you was worth a platoon, and three a company. The consolidation phase awaiting the inevitable counter-attack and, finally, the agonizing roll-call.

In these attacks a man's world was the shell-wracked horizon some few yards ahead. Under these conditions comradeship could help you carry on, but it was discipline that got you started and kept you going. And hanging over everything was that awful pretence of not being afraid.

Strange things happened in those trenches. There were those who said that some had a definite premonition of their death – a once-gay companion would fall silent and quietly go and prepare his kit; when the time came to 'pop the parapet' he would go mechanically as though his soul had already departed his body. Those ones seldom came back. Some other would get hallucinations and suddenly and for no apparent reason climb over the parapet to his death, unless his comrades could stop him in time. In the line they tried to maintain an offensive spirit in all they did, for the Irishman hates routine, but it was a lonely life. Yet none of this came out in their letters home – they lied in these, persistently and with intent, and pictured a life of comparative comfort and little danger.

Trench techniques were forever being perfected. The

night reliefs were the worst. The stealthy approach in pitch darkness across strange ground with grotesque shapes on every side and the dead glowing phosphorescent in the dark; the arrival in unfamiliar trenches, not knowing if that artillery concentration a while back had wiped out the leading men of the previous company; the hopeless congestion and seemingly total confusion as the new mingled with the old. The fervent hope that the enemy would not guess the hour of your relief this time. Those agonizing hours before dawn checking that you were in the right place, linked with your neighbours, and that no gap existed in the line; and prayer after heartfelt prayer that you would not be disturbed before the dawn when you could get your bearings by daylight.

Pervading everything was that sense of just surviving, without really living, under conditions of indescribable filth, among dead bodies, and in a general depression which seemed to have no end. Yet men, after a wound or spell of leave, would concoct any excuse to return to the front. There were absences or desertions, inevitably, yet it is a remarkable fact that the trenches and their danger and death possessed a bewitching fascination for the majority of those fighting in France.

For the young officers it was the experience of a lifetime, with undreamed-of responsibility at a very early age; to the men it was the overwhelming sense of comradeship in a unit as fierce in its pride as cheerful in its outlook. The confidence in their officers and in themselves was infectious in its enthusiasm and did much to sustain them in the bitter days that lay ahead.

The highlight of their spring stay in the rear areas took place in early March when the 2nd Battalion were withdrawn from the front and lay under canvas outside Calais. They held the Grand Calais First Spring Meeting on the sands outside the town. It was doubtful right from the 'off' whether or not the tide would turn in time, but all was well. Every soul in the Brigade – except the Church of England padre, who objected to the affair being held on a Sunday –

and quite a few outside it, turned out. Any animal on four sound legs was borrowed or commandeered, The fields in the five races were highly creditable, in number if not in quality. Anyone with any pretence at horsemanship rode, including Father Knapp of the battalion, aged fifty-eight. The battalion bookmaker, a newly joined ensign with an apparent flair for figures, was given a clerk and some cash and turned loose with instructions not to come back without a profit. Sporting the regulation brown bowler, a great furry coat and an artificial carnation in his buttonhole, he emerged comfortably in the black. St Patrick's Day followed soon after; the huge dinners and inter-company Gaelic football competition – a bloody affair if ever there was one – rounded off a memorable 'rest'.

In May they heard of the 'Easter Rising' in Dublin. There was no sympathy lost on the 'patriots', rather they despised them for not coming out and fighting a worthwhile enemy. Most Irishmen at the front felt a greater sense of betrayal than patriotism at their conduct – especially when many found their leave postponed as a result of the troubles.

So 1916 wore on. Both battalions were in that fearsome area known as the Salient. And it was in June they heard of the tragic death by drowning of their Colonel, Field-Marshal Earl Kitchener – Field-Marshal Sir John French (later the Earl of Ypres) was appointed in his place.

The Somme battles had started in July, with limited success. But the joint aims of relieving the pressure on the French, who were being 'bled white' at Verdun, and gradually wearing down the enemy's resistance in France and Belgium were being achieved – although it was proving a costly process, for one of the strongest places in the German line had been chosen.

'Not merely a series of successive lines, but one composite system of enormous depth,' so Sir Douglas Haig described the enemy positions. Every feature, natural or artificial, was turned into a fortress by all means known to the ingenious Germans. The bomb-proof shelters and deep communication trenches were the strongest the British had as yet encountered. The cellars which were to be found beneath every farmhouse were reinforced until they were veritable strong-points. The quarries and pits which abound in this country were buttressed and reinforced with concrete and made into dugouts for trench mortars or machine-guns. These were connected by underground passages, in some cases as much as thirty feet deep. The interlocking arcs of fire of more machine-guns swept across the open ground. There were even machine-gun emplacements sited so as to be able to cover the Germans' own trenches should these fall. In front of the whole, immensely strong defensive system were the wire entanglements. Two belts about forty yards in depth, built of iron stakes supporting barbed wire – not the agricultural type we know today, but with strands as thick as a man's finger. The Germans with justification believed the position impregnable – after all, it had been two years in preparation.

And this was only the first line of defence. The whole was repeated some 3,000 yards back and then again, and again, and again – the overall depth of this defensive position was little short of seven miles.

In conjunction with the French the first attacks on this formidable position went in on July 1st. By the 17th of the month the Germans had been forced back about three miles, at terrible cost. By September 12th, when the Guards Division were first engaged, a further 3,000 yards had been gained and Delville Wood, a prominent feature on this sector, was in our hands as were Ginchy and Leuze Wood, appropriately known as 'Lousy'. An ominous salient had been punched in the German line.

The Guards Division had been allowed to prepare for the coming offensive in the comparative peace of corps reserve. Every facet of the next battle was studied minutely and prepared to the finest detail. Rehearsals took place with the battalions formed up in lines moving forward behind the Corps of Drums, who were representing the creeping artillery barrage. A model was made of the exact area they were to cover at a scale of a foot to 100 yards, and it was of great value except that, as one survivor described it: 'It was like a doll's house garden and it looked you would be across it all in five minutes. But we was not, we was not.' Every

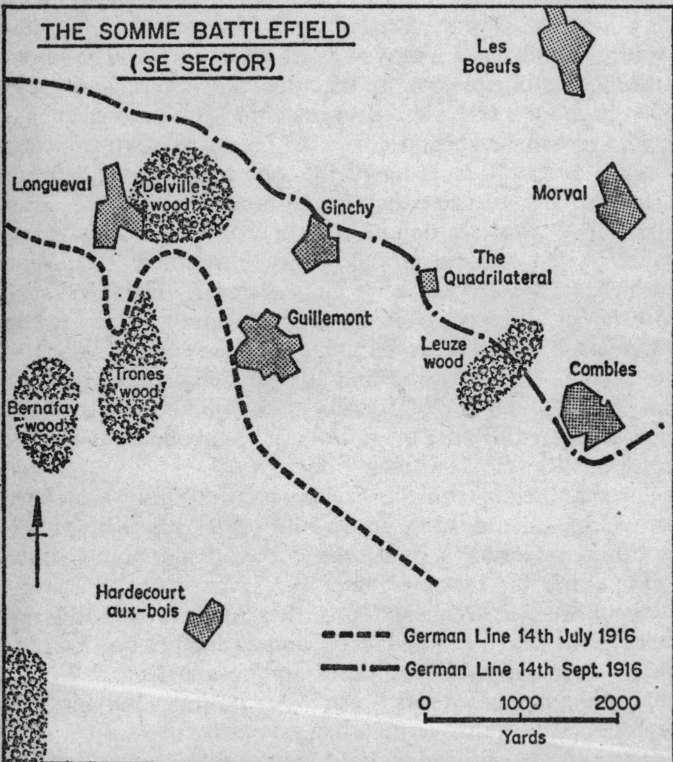

battalion, every brigade, had their own ideas on tactics and their own views on the ideal fighting order to be worn – these were all examined and paraded and discussed until a compromise, joint tactical doctrine was reached. The fact that this had been done and that there was now uniformity within the division greatly contributed to the success of their future operations.

In the midst of these hectic preparations they had their first sight of the 'creepy-crawlies', the 'apparitions', or as some called them the 'hush-hush birds' – the tanks which were to make their inauspicious début in this battle.

There were about thirty of them near the 2nd Battalion's camp. The Micks were not impressed. 'We had heard about moving forts that weighed thirty tons. Whatever it might be, we knew we'd have to take the thick of the coffee.' They were very right; zero hour for the attack had to be postponed for a full hour to give the tanks enough daylight to see where they were going – a delay which had a significant effect on the casualties in the 'Affair of Ginchy'.

It was to be a corps attack with the Guards Division in the centre. The front was only a thousand yards wide at the start-line – into this were to be crammed the 1st and 2nd Guards Brigades, each attacking with two battalions in the first wave. 3rd Guards Brigade was in close reserve, while the Cavalry Corps were poised to exploit should the opportunity offer – a situation they had enjoyed on many occasions.

The objectives were known respectively as Green, Brown, Blue and Red Lines, probing in all to a depth of some two miles into the strongest and best-prepared defences ever met.

The original start-line was an irregular series of trenches shaped in a rough semi-circle. It was to strengthen this that an attack was mounted by some companies of the 3rd Grenadiers and Number 2 Company of the 2nd Battalion Irish Guards. But caught in some uncut wire and enfiladed by some unsubdued machine-guns, the latter lost almost three-quarters of their small force within ten minutes of the start and were forced to withdraw. It was an unhappy prelude.

The country round about is described as 'undulating and gently rising to the east' to the heights around Les Boeufs. It was then absolutely flat except for the pock-marks of innumerable shell-holes. On their right flank lay the Quadrilateral – an awesome and intricate fortification which dominated the ground to the north and which was later to play a decisive role. On the left lay Delville Wood, as firmly in our hands as the lanes and ditches beyond it were not. These overlooked the rubble of Ginchy, and, incidentally, the left of the start-line. (See map, p 55.)

For the first time in their history three Coldstream battalions were to attack in line together, two in the first wave of 1st Guards Brigade on the left, and the other with the first wave of 2nd Guards Brigade on the right. The two Irish Guards battalions, also in action together for the first time, were in the second wave, neighbours in the centre; the 1st Battalion with 1st Guards Brigade, and the 2nd with 2nd Guards Brigade.

Just before the attack, Lord Cavan, the Corps Commander, sent the division a message: 'There are difficulties to be cleared upon the left and in front of 1st Guards Brigade and on the right of 2nd Guards Brigade, but the Commander-in-Chief is of the opinion that the situation is so favourable that every effort should be made to take advantage of it...' He could have added that they would be advancing uphill, across open ground dominated from a position outside the divisional boundary, from a start-line which was doubtfully secure and overlooked by the enemy!

Under cover of dark the brigades prepared for zero-hour. The two Irish battalions had come up as night fell, following the white-taped path laid for them, each man laden with two days' rations, a couple of bombs, extra bandoliers of ammunition, three sandbags and one or two other items in addition to their normal equipment – none of this much to the liking of the Irishmen, who prefer to fight light. They dug shallow trenches ten paces apart and which were no protection from the biting cold that night.

Shortly before dawn Father Browne, of the 1st Battalion, came forward and collected half companies or groups of men as he could find them. Together they knelt on the shell-churned ground, Protestant and Catholic, bare-headed, their rifles with bayonets fixed by their sides, while the small, seemingly insignificant figure of the priest gave them Absolution. This simple scene seemed somehow to throw into relief the squalor and the filth, the glory and the sacrifice which was their war. Father Browne said that it was the day set apart to Our Lady, and that not a few would be presenting their homage to Her in person ere many hours

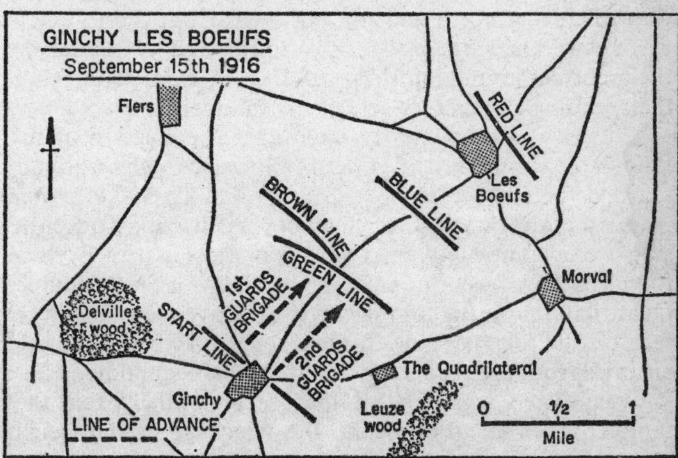

had passed. They returned to their scrapes silent and subdued, and there they waited.

Promptly at 6.20 AM on September 15th, 1916, the creeping barrage came down and the first wave of the two attacking brigades swept forward 'as steady as though they were walking down The Mall', described one eye-witness.

The Irish waited for their time and then followed, eager to close on the enemy before the original start-line came under fire. A couple of hundred yards out the leading battalions of 2nd Guards Brigade on the right came across a line of shell-holes full of fighting Germans. They were quickly subdued, but not before the following Irish and the rest of the second wave had stormed in on the heels of the first wave. This was the first confusion.

Worse followed almost immediately, for one of the leading battalions of the Coldstream over-ran their own creeping barrage, lost heavily in a cloud of bursting shells, and as a result started edging to the left. The tremendous fire from the Quadrilateral concentrated on the right-hand brigade and made the latter instinctively move away towards the left. The left-hand brigade itself, encroached on from the right, swung away in sympathy, and soon the whole line was advancing north and not north-east. But the following wave was unaware of this movement and continued on their correct course closing on those in front with whom their leading elements were already in touch.

Thus it was a hopelessly intermingled collection of battalions of both brigades and all waves which finally reached the first objective – Green Line. To make matters worse the attacks by the neighbouring divisions on either side were completely held up – so the Guards Division found itself with both flanks in the air, with an unsubdued Quadrilateral on the right, a body of active Germans in the rear on the left, heading in the wrong direction across the enemy's front and in a state of indescribable muddle.

Green Line was not all that secure either. But they tumbled over uncut wire into the wreckage of the trench, 'none too bad a refuge even if we had to bomb ourselves into it', they recalled. They then proceeded to fight and curse their way along. It was, in fact, rapidly becoming a general rendezvous – the bulk of the 2nd Irish plus a few of their 1st Battalion were there, as were some Grenadiers, the Scots Guards machine-guns and a mixture of all three Coldstream Battalions. Shortly after, the Commanding Officer of the 2nd Grenadiers, with some fifty men, and then Lieutenant-Colonel S. H. Godman of the 1st Scots Guards appeared. It was clear there was still work to do in the trench before it could be considered secure, so Colonel Godman ordered them to split and go right and left, clearing as they went.

More men came dribbling into Green Line 'like lost hounds, only they'd been fighting every yard of the way home'. In the confusion an officer grabbed whatever men he could find, irrespective of their regiment, and carried on with the business. When the officers fell, the sergeants took over; when they were killed the corporals assumed command, and when they died, as sooner or later they most of them did, the privates led on. The diary summing it all up merely said, with a wealth of meaning and not a little pride, 'the discipline held'.

Although the 1st Battalion Irish Guards had lost heavily on their way forward they had somehow kept out of the ruck – except for some men who had attached themselves to the 2nd Battalion and were now in Green Line already. But they, too, found it almost impossible to keep direction: 'We might have tried to guide ourselves by the waves of the sea, the way the shells spouted up.' They started to stray and ended up in another mess of the different regiments of their brigade, also in Green Line, but well to the north of where they should have been and separated initially from the main mass of 2nd Guards Brigade in Green Line – until such time as the two 'brigades' could clear the trench, join up, and make firm their conquest.

A hurried conference took place between Lieutenant-Colonel R. C. A. McCalmont, the Commanding Officer, and Lieutenant-Colonel J. V. Campbell, of the Coldstream, and then to the music of the latter's hunting horn the small combined force routed out the Germans on the next objective, Brown Line, in one screaming, cheering charge. Colonel Campbell won the Victoria Cross for his gallantry and inspired leadership in this action.

On the other brigade front Major C. E. A. S. Rocke, commanding the 2nd Battalion Irish Guards, Captain Sir Ian Colquhoun of the Scots Guards, and Captain O. Lyttelton (now Lord Chandos) of the Grenadiers, with a mixed force of some 120 men, had managed to push on to the third objective, Blue Line, just short of Les Boeufs itself, where they were astounded to find undisturbed ground and scrubby long grass, excellent cover for the Germans, who strongly resented their presence. And here it was that

Sergeant McLaughlin died. He was seen crawling forward on his hands and knees, or rather the stumps of his legs, for both had been blown off. 'I promised the Captain I would get to the final objective – I promised I would,' he kept muttering. He recognized an officer and grasped his hand. 'I promised I would get there,' he repeated once more, and then he died – but he had reached Blue Line. A very gallant gentleman, a Dubliner.

On Blue Line they dug in, sending back an urgent message for reinforcements which surprisingly was marked 'to be read by all officers on the way' – this can have done nothing to speed its progress. But none were available and the divisional commander, conscious of the failure of the two flanking divisions to advance at all, was disinclined to commit his reserve brigade to reinforce a position which was so precarious. So they extricated themselves with difficulty and fell back with surprisingly few casualties to rejoin their comrades also by now in Brown Line. And it was in Brown Line that the remnants of the two brigades stayed. They had nothing to do but, as one man put it, 'try to keep alive against the next day'.

The next day, September 16th, was little better. Shelled from dawn until dusk they hung on grimly. At one stage they watched an attack by a battalion of another division – saw it mounted; saw it falter; saw the few pitiful survivors drift back to the comparative safety of their own lines. That battalion attacked twice, on both occasions drawing the fire of all the enemy guns, even those who were shelling the Guards Division at the time. For a few minutes there was unbelievable calm then, as the attack failed, the shelling of Brown Line began anew.

They were relieved that night and eighty-eight officers and men of the 2nd Battalion Irish Guards and not all that many more of the 1st Battalion made their weary way back to breakfast at Bernafay Wood. Even over the span of years the veterans recall that morning in 1916. As Kipling wrote: 'The sun was shining, breakfast was ready for the officers and men near some trees. It struck their very tired apprehensions that there was an enormous amount of equipage and service for a very few men, and they noticed a sudden hustling off of unneeded plates and cups.' They ate and then they slept where they had eaten.

Survivors trickled in over the next days. After a time the flow slowed and then stopped altogether, and then came the awful realization that that was all who were to come back alive from the 'Affair of Ginchy'. But the Reserve Battalion had done their stuff again; the 1st Battalion received over 300 reinforcements and the 2nd Battalion a further 150 during the next days. Officers were scarce enough, but they could manage, just. When they were called upon to enter the arena once again six days later the 1st Battalion could muster 600 all ranks, and the 2nd Battalion 400 more or less fit men.

The intervening time was spent remaking and re-equipping the battalions. Some of them were employed in tidying up the litter of war behind the lines, where they were indiscriminately shelled. Good practice for the new arrivals, they reckoned, but the new boys stood the test. 'There was some wondering in the new drafts, but no budging, you'll understand.' However, there was work to be done and they were spared that terrible state of mind and body known as 'fighting Huns in your sleep'. On the 26th the 2nd Battalion took over. They in turn were relieved two days later and then the division was withdrawn to rest. Between them the two battalions of the Irish Guards had lost over 1,000 men in fifteen days.

The decorations started coming in later. They took the bravery for granted; and the credit they gave to their battalions. They looked at it with a philosophical eye. 'There was times when if a man was *not* earning VCs for hours on end he would not keep alive – and even then, unless the Saints looked after him, he'd likely be killed in the midst of it.'

And the news found its way back to Warley and the Guards Depot at Caterham, too. 'Yes. *We* made 'em – with the rheumatism on us and all; and we kept on making 'em till I got to hate the silly faces of 'em. And what did *we* get out of it? "Tell Warley that their last draft was damn rabbits and the Ensigns as bad." And after that, it's Military Crosses and DCMs for *our* damn rabbits.'

So 1916 extended into 1917 and they changed the mud of the Somme for the mud of the Third Battle of Ypres. As one put it: 'In those days, we had a grand choice of the frying-pan or the fire.'

Chapter 6

The Great War: Third Ypres to the Armistice

The situation at Verdun had changed radically at the end of 1916. In early November the French had achieved remarkable success on the Meuse, and subsequent operations in the same area consolidated these gains. There were signs of disintegration in the German battle array and the feeling was strong that one more push would see a major collapse in German strength and aspiration. The more so as the current German position after the Battle of the Somme was a particularly vulnerable one.

A major spring offensive had been planned, but before this could be mounted the Germans acted. Pulling back along an immense front, they withdrew to a new line of defence – the Hindenburg Line – which they had been preparing for months past. The BEF closed on the new line, but it was a gingerly performance in face of intense enemy artillery fire and across a landscape which had been razed to the ground and was bristling with booby-traps. But by early April the Allies were on the Hindenburg Line along its entire length.

In February 1917 the Germans had opened their campaign of unrestricted submarine warfare. Although this led directly to the entry of the United States into the war, it was also causing great inroads into our merchant shipping tonnage. To relieve this pressure it was decided to try again on the Flanders front with an ultimate aim of capturing the Channel ports. And so was conceived the Third Battle of Ypres which was to terminate in the slaughter that was Passchendaele.

The Guards Division had remained uninvolved throughout the early spring offensives but, on May 30th, they moved by train to the Flanders front. After lengthy preparations, including a lot of training for 'surprise situations', they moved into the line in the beginning of July.

The great air offensive in which over 500 British and 200 French aircraft took part began on July 11th. Partial mastery of the air had been gained by the end of the month, but not enough to guarantee that the occasional enemy aeroplane did not penetrate to the rear areas and make misery of the lives of those resting there. This was a new dimension in war and made 'rest' behind the front more of a misnomer than ever.

'In the old days, a trench was a trench and something to lay hold upon. Third Ypres was falling into nothing and then finding it was two pillboxes and a fort on your flank', was their condemnation of the new style fluid defence adopted by the enemy. But it wasn't enough to save the Germans on July 31st when the division advanced 2½ miles on a front of 1,500 yards – an almost unprecedented event in those days of trench slogging. Not without casualties, though, for the 2nd Battalion Irish Guards lost both Commanding Officer and priest.

The Commanding Officer was Lieutenant-Colonel E. B. Greer, MC, Alex's direct contemporary and fellow athlete. He had prepared his battalion for this battle with meticulous care and foresight. And the priest was dear old Father Knapp, close on sixty years of age and of indomitable courage – he was loved not only by the 2nd Battalion but by the whole brigade, to whom he was invariably 'Father'. A familiar sight in the front line, with white hair flowing in the wind and a blackthorn stick in his hand. Before an attack he would walk up and down the trenches asking his men if they were at peace. His presence had an incalculable effect, especially on the younger soldiers who now made up the vast majority of the battalion. After the attack he was to be found wandering behind the front line calling out to the men, seeking the wounded who might have been missed by the stretcher-bearers. Many a Mick owed his life to the fact that Father Knapp never gave up his painstaking search after a battle. He never seemed to rest.

Then came the rain. Four days of it. Shell-holes became lakes, trenches became rivers, work was almost impossible but they did it, if only to keep themselves occupied. The existing trenches were totally uninhabitable, so they dug more. When these became unusable, as they did after a couple of days, they carved out another line. In the midst of it all the division was moved again – to another, and if possible wetter, part of the Flanders front: on the Broenbeck River, near Langemarck, south of the Forest of Houthulst and some six or so miles north of the site of the unhappy but gallant operation around Polygon Wood in 1914. (See map, p 32.) But as they moved, so the weather cleared.

The Broenbeek was a nasty little stream, varying in width from eight to twenty feet. The water was only some three feet deep, but beneath it lay a bottomless layer of mud, and the whole was liberally strewn with barbed wire. The approaches were equally uninviting and in full view of the enemy.

It was this stream which divided the outpost company of the 2nd Battalion Irish Guards from the remainder of the battalion. The platoons had hardly moved in when a generous barrage broke about their ears; as it lifted, some 200 German infantry (they turned out to be Würtemburgers) wearing body-armour rushed those who were left after the shelling.

After that débâcle the folly of allowing the stream to split the forward troops was acknowledged and the battalion was allowed to take up its position behind it. Patrols were sent to look for any survivors, but few were found and it was clear that the Germans were now in firm control of the little river. Then late in the afternoon of the following day five SOS rockets were fired by someone off to the left followed by an earth-shaking barrage from both sides on the edges of the river. Loath to be left out, the machine-guns joined in to swell the din. It was impossible that anything could live through such fire – or so they thought. At last the racket died down and then, some two hours later, in walked a complete missing platoon. They were hungry and very tired, but in good order and complete.

It transpired that their outpost position had been completely cut off by the Würtemburgers' raid, and when that finished they found a machine-gun had been set up between them and the river. Sergeant Moyney, who was in command, thereupon ordered them to remain quiet. They had a bottle of water apiece and some iron rations, but he had been instructed to stay put and leave the rations intact, so these were untouched. Kipling takes up the tale: 'The enemy knew they were somewhere about, but not their strength nor their precise position, and having his own troubles in other directions, it was not till the dawn of the 16th that he sent out a full company to roll them up. The Sergeant allowed them to get within twenty-five yards and then ordered his men to "jump out and attack". It was quite a success. Their Lewis-gun came into action on their flank, and got off three drums into the brown of the host while the infantry expended four boxes of bombs at close quarters. Sergeant Moyney then gave the order to charge through the Germans to the Broenbeek. It was done, and he sent his men across that foul water, bottomed here with curly barbed-wire coils, while he covered their passage with his one rifle. They were bombed and machine-gunned as they floundered over the swampy western bank; and it was here that Private Woodcock, hearing cries for help behind him, turned, waded into the water under the bombs and bullets, fished out Private Hilley, of Number 3 Company, with a broken thigh and brought him to safety...' Both Sergeant Moyney and Private Woodcock were awarded the Victoria Cross for their conspicuous gallantry. Private Woodcock was killed in action in March 1918; Sergeant Moyney is alive today and lives in Roscrea, County Tipperary.

The 2nd Battalion was invited to recross the Broenbeek on October 9th. A two-brigade attack preceded by a massive barrage was wholly successful and brought an end to their present labours and for the Irish Guards their contribution to the Third Battle of Ypres.

So their tour of Flanders ran to its noisy, weary, muddy conclusion and, on November 11th, 1917, after three weeks' rest, they moved back to the Somme. The Battle of Cambrai started nine days later. (See map, p 66.)

On the sector to the west of Cambrai, the first German defensive system, the Hindenburg Line proper, ran from Banteux on the Scheldt, north-west to Havrincourt, and then north along the Canal du Nord to Moeuvres. To the east lay the next position, which followed roughly the line Crêvecoeur–Marcoing–Fontaine Notre-Dame–Bourlon Marquion. The key to the whole structure on this front was the high ground around Bourlon, and it was this ridge which was to prove the one flaw in the resounding success of the Battle of Cambrai.

For the first time tanks were to be used en masse rather than in the 'penny-packets' so derided by the experts. There was no preliminary bombardment; surprise was complete. By the time evening closed on the first day the advance was three miles deep. The next day saw the capture of the Flesquières Ridge, but still Bourlon Ridge remained in German hands. Sir Douglas Haig had reckoned that he had forty-eight hours to achieve his objectives before German reinforcements could reach the threatened sector. This time was nearly up. But until Bourlon Ridge had fallen no further advance was possible. Attack after attack failed, or succeeded only momentarily, until the Germans counter-attacked and drove the attackers out.

Two battalions from 1st Guards Brigade were borrowed by the hard-pressed 40th Division, who had been struggling to gain and keep a foothold. At midnight the company commanders of the 2nd Battalion Irish Guards were told that they were to attack at 5.30 AM on the following day.

The importance of the ridge with its wood was obvious to everyone, not least the Germans, who had recovered from the initial shock of the battle and had had ample time over the last days to develop their positions. Further, although it wasn't known at the time, the enemy were planning to counter-attack that very day on the same sector and had massed their reserves in preparation for it.

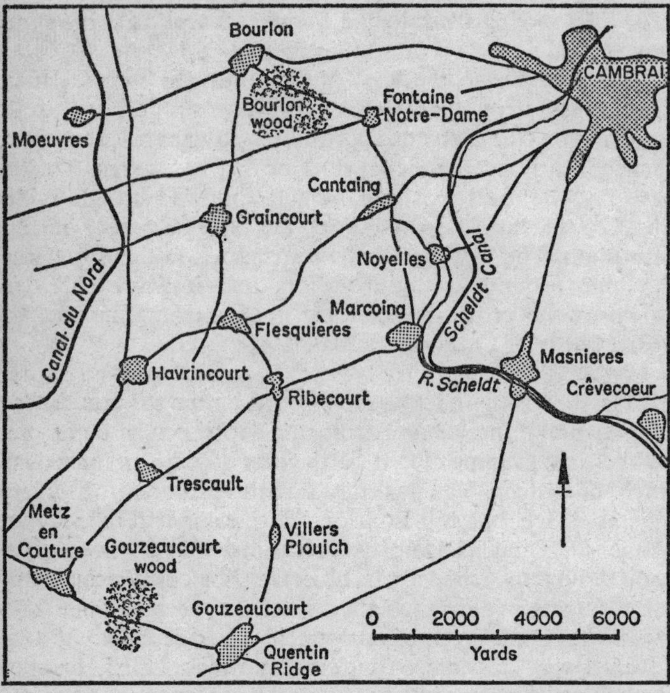

The result was a foregone conclusion – but, as so often happened, all seemed to be going well at first. Promptly at 5.30 AM the barrage came down and they crossed the start-line which was in the wood itself where the thick undergrowth hampered their movement and made control almost impossible. The few tanks which had been allotted to this attack arrived late and took no part. With great gallantry the three battalions of 2nd Guards Brigade fought their way forward against tremendous machine-gun fire and intensive enemy shelling. But their own barrage went too quickly and within minutes the two leading companies of the 3rd Grenadiers on the right had lost all their senior ranks except one NCO; in the centre, the Coldstream were in little better shape but had managed to secure their objectives with great dash; the Irish on the left, who were supposed to link with a parallel attack by the 186th Infantry Brigade, were badly caught in the flank by well-sited machine-guns in the eastern end of Bourlon Wood.

Direction-finding was impossible amongst the heavy undergrowth in the wood. The ground was waterlogged and as sticky as any they had come across. The rides in the wood were as misleading as their 1st Battalion had found those at Villers-Cottérêts, and here was the bonus of heavy, accurate and persistent shelling – for the Germans had been happily ranging on this area for the past six days and the whole defence of the place had been planned with meticulous efficiency.

The impetus of the brigade attack faltered and stuck. Their objectives had been secured but they were far too thin on the ground either to hold their gains or sweep the area they had won. They could hardly find enough guards for the prisoners, of which they had taken a good number, so the latter seized every opportunity to bolt away and rejoin their comrades who had been by-passed and were lurking waiting for the moment to strike. Some, when they escaped, also took up rifles from the fallen and attacked the battalions in the rear. The cry for urgent reinforcements went back, but by the time any could arrive it was too late, and the Germans were methodically shelling behind the leading troops to prevent those reinforcements getting forward.

When the inevitable counter-attack came at last the Germans found the gaps in the line with little difficulty, and the remnants of the brigade had to pull back, sharply.

The whole of the 2nd Battalion was all but cut off. Splitting up into small parties they carved their way back. At noon Battalion Headquarters, where the Commanding

Officer, Lieutenant-Colonel Alexander, remained as imperturbable as ever despite the catastrophe which was looming nearer and nearer, confessed to having lost all touch with the battalion. It looked like complete disaster. And it would have been, said one, 'If Jerry had only shut off his damn guns and *listened* he'd have heard we was knocking about; but he kept on hammering and rushin' his parties back and forth through the wood, and so, you see, them that could of us, slipped back quiet in the height of the noise.' The diary said that they fell back to their original line with the rest of the brigade but, unofficially, 'we did – but I don't know how we did it. There wasn't any battalion worth mentioning when we were relieved in the dark, but stray men kept on catching up all night long.'

Some eighty all ranks came out of that wood that day – over 400 had gone in. Battalions in 1917 went into battle with only three officers per company and platoons at 50 per cent strength; the remainder were left out of battle to act as nucleus for the reinforcements for the next contest – it was as well they did. But such was the strength of the reinforcement reserves that the War Diary could blandly record that four days later the 2nd Battalion were 400 men strong and once more a viable force.

The Germans had conclusively won the battle of Bourlon Ridge; now it was their turn to attack. As the British momentum faltered and died the enemy struck. At breakfast time on November 30th they launched themselves at what we called a salient and they called a bag.

In the north they were held by the stubbornness of three divisions and the bravery of some gunners firing over open sights, but to the south the situation was different. In a matter of hours, and despite a resolute defence, that flank was in ruins. By noon the enemy had swept on for over three miles, recapturing all the ground they had lost during the previous weeks, and even the original British line itself. Disaster seemed imminent.

By the greatest good fortune the Guards Division had been pulled back out of the line to Metz-en-Couture, a few miles from Gouzeaucourt (see map, p 66), where they were licking their wounds and reforming after their efforts at Bourlon. It was a time of horrible conditions, of frost and snowy rain. But they found themselves full in the path of the German advance.

The roads were thronged with troops hell-bent for safety. Many were without their weapons or equipment, and it was clear the panic was general. Horses, some riderless, wagons, and ambulances swept past them, even some gunners, still faithfully clutching their gun-sights.

The situation was critical and 1st Guards Brigade were committed straight from the line of march without a pause. There was no artillery support – their own guns were miles to the north. They halted and fixed bayonets outside Gouzeaucourt Wood and went straight into the attack. The three battalions reached the top of the slope above Gouzeaucourt almost simultaneously, having suffered remarkably few casualties, and then they swept into the village and on to the ridge beyond, in an irresistible charge.

The Irish picked up on the way some sappers, both British and American, who were still gallantly holding out in a trench, and some dismounted cavalry, who appeared from nowhere, joined in. The enemy were seen to be leaving in some hurry and it was warm work catching up to bayonet them.

The next day 3rd Guards Brigade, with strong support from some hurriedly assembled artillery, mounted a successful attack on the Quentin Ridge beyond the village where the enemy had been shelling them with captured weapons as well as their own. With the ridge in our hands the position was saved, and any counter-attack was swallowed up in the snowstorms which then followed. So stalemate returned.

The happenings in November 1917 were but a microcosm of what was to happen the following spring. Tour had succeeded tour in the front line. Patrols were active, the occasional raid took place, shelling varied from spasmodic to intense, casualties continued without cease. But 'we could *feel* Jerry piling up against us in those days'; so they described the unreality of the glorious spring weather of 1918.

On the morning of March 21st, under cover of a dense fog and a bombardment of unexampled severity, the promised attack broke. Along a front of fifty miles the Germans stormed forward. Within a short time most of the advanced defences had been over-run, and in the south towards Flesquières the situation was critical with all local reserves committed. The Guards Division were thrown into the fray.

As the division set about digging a new line, the transport was sent back out of harm's way. Some news filtered in: Gouzeaucourt, of treasured memory, had fallen; so had Péronne to the south; Amiens itself was threatened; it seemed as though the whole front was crumbling. More alarming still were the now growing number of stories of rout and disorder.

April 9th, for 4th Guards Brigade, in which were the 2nd Battalion Irish Guards, was to have seen a rehearsal of an inspection by their Divisional Commander. The cynics among them voiced a general misgiving: 'When there's ceremonial after rest and fat-up, it means the General tells you all you are a set of heroes, and you've done miracles and it will break his old, hard, heart to lose you; and you'll trot off at once, up the road, and do it all again.' So when the following afternoon they were told to proceed to an unknown destination, they were little surprised.

They jolted northwards by bus for the whole of April 11th, arriving late in the evening after a most unpleasant journey. It was a strange land and bitterly cold. The Commanding Officer went off to find the Brigade Commander, for there was no sign of the other battalions. And so, after posting sentries, 'to keep a good look-out, in case we find ourselves in the front line', a course which looked more and more probable as the night wore on, they settled down to a chilly wait.

Those hard-pressed British and Allied divisions trying to hold back the German flood in the Merville area had fought gallantly against great odds, but it was only a question of time before they must be overwhelmed. Friday, April 12th, was the crucial day. By this time the steady German pressure was beginning to tell. The few reinforcements that

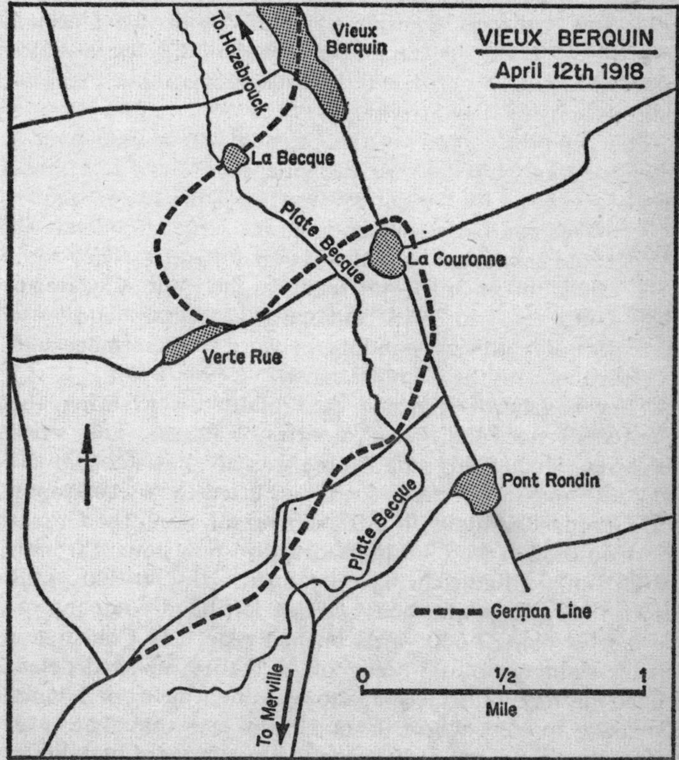

could be spared were on their way or already committed, but they were little enough as the rest of the British front was in almost as parlous a state. However, a fresh division, the 1st Australian, was due to detrain at Hazebrouck within thirty-six hours; at all costs this time must be bought and the line held intact until then.

This was the situation which faced 4th Guards Brigade

when they arrived at Hazebrouck. By dawn they were more or less complete, and were thrown directly into the fight.

It was a flat, prosperous-looking countryside they were in, unspoilt by war and quite open; a startling contrast to the mud and trenches they had grown accustomed to. They could see balloons floating peacefully over the German lines – but it soon became apparent that they were under observation, too, for a heavy barrage broke about them almost as soon as they arrived.

The position chosen by the brigade was a little over a mile north of Merville near the village of Vieux Berquin – partly protected by the narrow stream of the Plate Becque. It soon became clear that both flanks were completely in the air and that there was no support for miles. They took up a line with the 3rd Coldstream on the right, 4th Grenadiers on the left and 2nd Irish Guards in reserve. And those were the only troops standing between the Germans and Hazebrouck and the Channel coast.

They had hardly reached their allotted place when the Germans started the first of a series of attacks. They must have sensed that this small force was all that stood in the way of a massive victory, for they hurled themselves upon 4th Guards Brigade with a ferocious zeal. Soon the Grenadiers on the left were taking a dreadful punishment from a battery of field-guns firing over open sights at 300 yards down the road, while their extreme left-hand company reported Germans 1,000 yards in their rear. The Coldstream on the right were little better off, and they reported determined efforts by the Germans to get round their open flank. To make matters worse it soon transpired that there was also a gap in the centre between the two forward battalions.

All the Brigade Commander could do was to send a company of Irish Guards to each of the forward battalions. So Number 3 Company were dispatched to the Grenadiers, with whom they duly suffered, while Number 2 Company went to the Coldstream.

The defenders were hampered in their efforts by the stragglers who strayed past them from time to time. There seems to have been no hysteria: 'We told them to stop and help us. Lots of 'em did. No, they didn't panic a bit. They just seemed to have chucked it quietly.'

As dusk fell the brigade were still holding firm, but no one quite knew where the line lay. The remaining companies of the 2nd Battalion had been committed one by one as occasion offered, and Battalion Headquarters itself had been forced to move three times.

By this time the whole brigade seems to have been more or less surrounded, but there is no coherent account of the battle. The night was a restless one, and as a misty dawn broke on April 13th, the critical moment arrived. Critical alike in this battle and, in retrospect, for the whole German offensive on that front.

The crisis was upon them almost immediately. An urgent message came through that the Germans had at last broken in between the Coldstream and the Grenadiers. Number 3 Company were instantly dispatched and vanished into the morning, never to be seen again. Shortly after, another call came that the extreme left of the line was surrounded and in imminent danger of collapse, and thither went another company of the Irish; of that party only one NCO and six men came back late that night.

4th Guards Brigade now consisted of a number of small isolated groups, surrounded by an overwhelming mass of Germans. The epic stand of Captain T. T. Pryce, commanding the left-flank company of the Grenadiers, was typical of the combat that day. Early on the 13th he reported his company was completely surrounded, his numbers reduced to thirty men and that he was running out of ammunition. The survivors fought on all day standing back to back – by evening their ammunition was gone, but they still had their bayonets, and when the Germans next attacked, they charged; there were only eighteen of them left by this time. The enemy couldn't fire for fear of hitting their own men behind the small band. But the pitiful size of the remnant was now clear and when Pryce and his party, now only fourteen strong, charged again, it was too much: the gallant Captain and the few men left of Number 2 Company the 4th Battalion Grenadier Guards were swallowed up and annihilated. Only one man survived to tell of this incredible action. Captain Pryce was awarded a posthumous VC.

Elsewhere along the brigade's 4,000-yard front the resistance continued. At one time the Headquarters of the Coldstream and Grenadier battalions found themselves, somehow, fighting side by side and beating off the enemy who were behind them in great strength. The small groups held on; no one to this day knows quite how. The execution they committed was fearful. Throughout, the steady, disciplined rifle fire, which the Germans still considered was from machine-guns, held the enemy at bay. As long as the ammunition lasted, the Germans hesitated from closing, but then, when all bullets were spent, the few survivors of each little party were at the full mercy of the enemy masses. As darkness fell the crisis was over, however, for the Australian division had now arrived and the way to Hazebrouck was barred to the Germans, for ever, as it happened.

The brigade received praise in plenty for their wonderful effort. Sir Douglas Haig wrote: 'The performance of all troops engaged in this most gallant stand, and especially that of 4th Guards Brigade, on whose front of some 4,000 yards the heaviest attacks fell, is worthy of the highest praise. No more brilliant exploit has taken place since the opening of the enemy's offensive, though gallant actions have been without number.' This was not much consolation for their crippling losses. The 4th Battalion Grenadier Guards lost fifteen officers and 504 men; the 3rd Battalion Coldstream Guards also fifteen officers and 490 men; and the 2nd Battalion Irish Guards nine officers and 250 men.

That was the end of 4th Guards Brigade as a fighting force. The Grenadier and Coldstream remnants were absorbed into a composite battalion; it was not worth combining the survivors of the Irish Guards with anyone and the 2nd Battalion, which had been under strength before the battle started, from that moment only existed to keep the 1st Battalion operational, though they were not actually disbanded until 1919.

But shattered as they had been at Vieux Berquin, what turned out to be almost the last parade of the 2nd Battalion Irish Guards in the war showed that their spirit was still alive. On May 15th, 1918, they were inspected by General Plumer for a distribution of medal ribbons. The rehearsal the day before had been thoroughly bad, but on the day itself they achieved on parade a faultless ceremonial performance – drill, turn-out and appearance all excellent: 'The truth is,' described one, 'the way we were put through it at Warley, we knew *that* business blind, drunk or asleep, when it come to the day. But them damn rehearsals, with the whole world and all the young officers panicking, they're no refreshment to drilled men' – a sentiment that is sometimes echoed today.

Yet the end was near. The German offensive was spent and the rot about to set in. May and June were quiet months for the Guards Division and in July the bill of casualties was one man killed and two wounded, the latter when the Germans welcomed the arrival of the Americans with a barrage.

The severe losses of the spring had been made good by fresh drafts from home and a thinning out of secondary theatres. The BEF was now as strong as it had ever been, stronger, indeed, in artillery. On August 8th the expected attack broke upon the Germans in the Amiens area; it was perfectly concealed and the support of 400 tanks was decisive. By the end of the month the constant eroding of the German position was beginning to tell. There was an unmistakable feeling of elation in the air. The enemy still fought stubbornly but enemy prisoners now seemed thankful to be captured.

The Germans were pulling back all along the front – to the Hindenburg Line. The breaking of this was to be one of the bloodiest of all the bloody battles in the war. 'And we knowing we was told off to cross that dirty ditch in front of 'em all' was how they put it.

The dirty ditch referred to was the Canal du Nord (see map, p 66) which stretched right across their front, screening the mass of the Hindenburg Line. The battalions had had time and enough to ponder this operation and they knew the ground well; nevertheless, it was a formidable undertaking that lay ahead.

2nd Guards Brigade were to launch the initial assault, cross the canal and carry on to a depth of about 2,000 yards; when firm, 1st Guards Brigade were to take up the running, for a further 3,000 yards; and then 3rd Guards Brigade were to pass through and assume the lead.

The 1st Battalion Irish Guards of 1st Guards Brigade moved forward from the reserve position punctually at 3.30 AM on September 21st, 1918. No sooner were they in the battle area than up went the tell-tale signal from the German lines, two red lights above one below. 'Up go the bloody pawnbrokers,' was the cry; and down came a barrage from the trench-mortars. They plugged on to their allotted position some fifty yards in rear of the 1st Scots Guards who were the advance of the leading brigade.

Zero-hour was at 5.20, and dead on time a fearsome barrage came down on the luckless defenders. Promptly the Scots Guards moved off, followed by the Irish a few minutes later. The bed of the canal was dry and concrete hard. They saw the Scots Guards seemingly tiny in the half-light mounting scaling ladders against the far bank and scrambling up. The canal bed was full of Germans clamouring to surrender and pathetic in their eagerness. 'We'd have laughed if we had had the time,' said one private.

They pressed on and by seven o'clock the Scots Guards were firm on their objective and 1st Guards Brigade passed through without incident and attacked the next line.

Their progress was slow but steady. Already they were being impeded by prisoners who were becoming an embarrassment as one officer found when a youth, who cannot have been yet seventeen, flung himself into his arms shouting, *'Herr Offizier! Ich bien sehr jung! Kamerad!'* For which he got scant sympathy. 'Get on with you. I wouldn't touch you for all the world.'

The whole place was one huge Hun-warren. There were Germans everywhere, a few with fight still in them. But by 11 AM the brigade were firmly established and 3rd Guards Brigade passed through, shortly afterwards followed by 2nd Division, for there was to be none of that hiatus between follow-up formations that had bedevilled the chance of their 2nd Battalion at Bourlon Wood. It was a copy-book operation throughout. They lost fifty per cent of their officers and forty per cent of their men, but it was a winning battle and the way was now clear.

'An Armistice was declared at eleven o'clock this morning, November the 11th,' was how they heard it was all over. Kipling telling of the next few hours writes, 'Men took the news according to their natures. Indurated pessimists, after proving it was a lie, said it would be but an interlude. Others retired into themselves as though they had been shot, or went stiffly off about the meticulous execution of some trumpery detail of kit-cleaning. Some turned round and fell asleep then and there; and a few lost all holds for a while. It was the appalling new silence of things that soothed and unsettled them in turn. They did not realize until all sounds of their trade ceased, and the stillness stung their ears as soda-water stings on the palate, how entirely these had been part of their strained bodies and souls. It felt like falling through into nothing. Listening for what wasn't there, and trying not to shout when you remembered for why.'

They entered Germany at the beginning of December to the music of their pipes and drums. And then the rain came to add finish to it all. And 'we was *all* wet most of the time and fighting with the mud and our boots. It was Jerry's own weather the minute we set foot in his country, and we none of us felt like conquerors. We was just dripping Micks.'

Christmas 1918: 'They sat them down, twenty-two officers and six hundred and twenty-eight other ranks, and none will know till Judgement Day how many ghosts were also present. For the first time since August 1914 the monthly returns showed no officer or man killed, wounded or missing.'

The Regiment had been awarded four Victoria Crosses (including one to Lieutenant-Colonel J. N. Marshal who was serving with the Lancashire Fusiliers at the time); fourteen DSOs and seventy-seven DCMs. They had lost in all 2,349 dead, including 115 officers; and their total of wounded was 5,736.

'But you'll understand, when everything was said and done, there was nothing *real* to it, except when we got talking and passing round the names of them we wished was with us. We was lonely in those days. The half of us was Church of England by then too. But we were lonely, you'll understand, as units. And our billets, mind you, ma-agnificent, with walls and lockers and doors and all. The same for the officers! And there was Mr — that I'd known well any time these last two winters, freezing and swearing alongside of me in any shell-hole we could find, and glad to be out of the wind – and now him cursing in his quarters because he had not the Jerry-talk for the German for: "Turn off that damn steam-heat!" And that's War also. But you might tell we was lonely, most of all. Before God, we Micks was lonely.'

And that is, perhaps, the most fitting epitaph of all.

Chapter 7

Between the wars

The months waiting to get home seemed interminable. By the end of 1918 the two battalions of the Irish Guards were at Cologne as part of the Army of Occupation and the Brigade of Guards was fast reassuming its customary parade ground appearance.

In January the Prince of Wales presented Colours to the battalions of the Brigade which had been formed during the war. It was a sad as well as a proud moment, for the men of those battalions knew that in a few months their units would be no more and that the traditions they and their past comrades had carved and created would fade into history. At the same time it was announced that in honour of their gallantry in the war privates in the Brigade of Guards would henceforth be known as Guardsmen.

At last, in the spring of 1919, the Guards Division came home; and on March 22nd London paid tribute to her Guards. For the Guards belong to London, and in those days even in the Irish Guards there were a good few London recruits.

It was a great occasion for all who took part and for the schoolboys from the Officers' Training Corps who lined the streets. The whole of London seemed to have turned out for the homecoming of their Guards. The Scots Guards brought along the two cows – Bella and Bertha – which they had acquired at Neuve Chapelle, and for the first time two Irish wolfhounds belonging to the two battalions of the Irish Guards were on parade together. One of them caused a roar of laughter when, pressed by the call of nature, he towed his diminutive drummerboy keeper to the nearest lamp-post.

Demobilization proceeded slowly but steadily. Unfortunately the civilian labour market was quite unable to absorb all the ex-soldiers and many, when the promised release came at last, found that they were hard put to adapt themselves to civilian life.

After four shattering years of war it was difficult for many, impossible for some, to take up the threads of their lives again. It was a drastic readjustment that they had to make. Some were destined never to recover from it. Young in years but prematurely aged by the horrors they had witnessed and experienced, they moved shiftlessly from job to job, incapable of settling to any routine and loathing the peace, and, for some, the equalization of civilian life. Quite a number became nothing but empty shell-shocked alcoholics, able only to recount the tales of their war, which soon became in their mouths glorious in the telling; and for many others the period of change was long and trying.

To help in this process of adjustment the Regiment founded the Old Comrades' Association – now called the Irish Guards Association and currently with branches in Dublin, Belfast, Liverpool, London, Birmingham and Manchester. It is the function of the Association to find employment for those Irish Guardsmen who have left the Colours and to ensure that their own and their dependants' subsequent welfare in civilian life is properly supervised.

As the Army returned to its peacetime form many regiments had to be disbanded, and on March 31st, 1919, the 2nd Battalion Irish Guards were put into suspended animation and their Colours laid up in the Roman Catholic church at the Guards Depot at Caterham. The 1st Battalion were then stationed at Warley.

That battalion now resumed its peacetime duties of drill and field training, but the field training had a more purposeful look about it than in the far-off days before the Great War. On St Patrick's Day 1920 the Guards returned to Full Dress, and to many that was the outward symbol that the war was really over. On St Patrick's Day the following year the Pipers of the Regiment wore their full dress of saffron kilt and green cloak for the first time.

The promise of a humdrum peacetime existence would not satisfy a number of officers, and they sought excitement elsewhere. Among these were Lieutenant-Colonel Alexander and Captain J. S. N. FitzGerald who found occupation in the Baltic States in the war against the Russians.

Having lost their 2nd Battalion, the Irish Guards assumed that the ravages of peacetime retrenchment would leave them untouched, but the following year a far graver menace appeared; one which threatened the very existence of the Regiment itself.

The immediate post-war disbandments had drastically cut down the military establishment of regular infantry battalions. But this retrenchment proved inadequate, and the expense of maintaining so large a standing army was too great; more had to go. The fact that further cuts were required was well enough known. But the Brigade of Guards felt themselves immune. They were to be rudely disillusioned, for the two most recently formed regular regiments in the British Army came under the butchers' eyes; and the Irish and Welsh Guards found themselves fighting for their very existence.

The proposal was to cut the Welsh Guards to one company which was to be incorporated into the Grenadiers; while the Irish were to lose their separate identity and their Regimental Headquarters and to become in effect a third battalion of the Scots Guards. To the planners this produced the orderly situation of three Regiments of Guards each of three battalions.

But the proponents of the scheme can never have anticipated the storm of protest that arose when it became public. The matter was still under discussion in the War Office when an ex-Irish Guards officer, Valentine Williams, then working for the *Daily Mail*, got wind of the proposal long before it was due to be announced. On June 4th, 1920, he broke the news. Outraged protests rose from every side, not least from the powerful lobby of Welsh Members of Parliament who took the proposed reduction of the Welsh Guards as a deliberate affront to the Principality. The arguments were thrashed out in public and in private and the Irish Guards were fortunate to have such a formidable champion as their Colonel, Lord French, who was then Viceroy and Lord-Lieutenant of Ireland. The matter was submitted to the King; and in the House of Commons, on June 15th, 1920, Mr Winston Churchill was able to announce: 'This subject has now been considered by the Army Council. There is no intention of disbanding the Irish or Welsh Guards so long as they are able to maintain their recruiting in such a manner as to preserve the national character of the Regiments.'*

The crisis was over. And for the Irish Guards it was fortunate that the matter was so thoroughly discussed in 1920, for in 1922 the Troubles in Ireland ended, the Irish Free State was created and the disbandment of the Southern Irish regiments became inevitable.

On June 12th, 1922, at Windsor Castle, the Colours of the Royal Irish Regiment, the Connaught Rangers, the Leinster Regiment, the Royal Munster Fusiliers, and the Royal Dublin Fusiliers, all regiments with reputations second to none for courage and tenacity in the field, were handed over for safe-keeping. On July 31st, 1922, those Irish regiments officially ceased to exist. With their departure from the roll-call of the British Army ended an era; it was a bitter moment. Their regimental silver, some of it of priceless value and of great historical worth, passed to the remaining Irish Infantry Regiments, the Royal Inniskilling Fusiliers, the Royal Ulster Rifles, the Royal Irish Fusiliers and the Irish Guards.

Much has been written about the tragic events which occurred in Ireland from the time of the Easter Rising in 1916 until the signing of the Treaty in 1922. It is not intended here to elaborate on all that took place during those sad years,

* Official Report, House of Commons, Vol 130, Col 1061.

but it is an incredible fact that the allegiance and loyalty of the many Irishmen fighting in the British Services were not more affected by the strife which was taking place in their beloved country, and by the political issues involved.

Shortly before the disbandment of the Irish regiments the Irish Guards received orders to join the British garrison in Constantinople. The order to move was received early in March and, as embarkation leave was granted to all ranks, the St Patrick's Day parade had to be abandoned for the first, and so far the last, occasion in peacetime. On Friday April 21st, 1922, the King inspected the battalion and four days later the Irish Guards embarked on HMT *Derbyshire* for their first peacetime tour abroad. Lieutenant-Colonel the Hon T. E. Vesey, who had seen the battalion through the difficult immediate post-war readjustments, had given up command and been succeeded by Lieutenant-Colonel Alexander, recently returned from service in the Baltic States.

It was of Colonel Alexander that Kipling wrote when Alex gave up command of the 2nd Battalion Irish Guards in early 1918: 'At the worst crisis he was both inventive and cordial and, on such occasions as they all strove together in the gates of Death, would somehow contrive to dress the affair as high comedy. Moreover, when the blame for some incident of battle or fatigue was his, he confessed and took it upon his own shoulders in the presence of all. Consequently, his subordinates loved him, even when he fell upon them blisteringly for their shortcomings; and his men were all his own.' He loved his Micks and his Micks loved him. At this crucial time in Irish affairs there could have been no happier choice for a battalion, many of whom were deeply troubled by what was happening in their own country. Alex's superb leadership and light touch, together with the magic of his personality and unrivalled reputation as a fighting Mick, held the battalion together at a difficult time.

After a short pause in Alexandria, the *Derbyshire* passed through the Dardanelles and reached Constantinople at daybreak on May 12th. With colours flying and bayonets fixed the battalion marched to their barracks in Pera, the European quarter. Sited high over the Sea of Marmora, with a magnificent view of Asia Minor across the Bosphorus, Tash Kishla Barracks, as it was called, was to be their home for nearly a year.

The 1st Battalion Irish Guards were now with the Army of the Black Sea, commanded by Lieutenant-General Sir Charles 'Tim' Harington, the British contribution to the Allied force which was designed to implement the post-war policy of keeping the peace in the Near East, and was positioned in Constantinople with the specific task of ensuring that the Straits remained open to international shipping. In Turkey itself the aged Sultan, confined for his own safety to Constantinople, could call only upon his immediate bodyguard and household troops for support; the rest of the country were firmly behind Kemal Ataturk who was fighting a gallant and a winning war against the Greeks in Anatolia.

To give teeth to the Allied diplomatic argument the three nations could muster eleven battalions – four British, six French and one Italian. A total of 7,600 men against the Greeks or Turks, each of whom could muster ten times that number. Of greater significance were the six capital ships of the Mediterranean Fleet.

With such a small British contingent in Constantinople the arrival of the Irish Guards was more than welcome, and soon the battalion settled down to their role of garrisoning the city. Although the majority had seen war service, this was the first time that all but a handful had ever served in a Middle Eastern country. The habits and pleasures of the East fascinated them and they were not long in discovering the hidden potency of the local drink – 'dusico' they called it – which could leave a man sober in the evening and a raving drunk after consuming his tea at breakfast the following morning. Mosquito nets were an entirely new feature, and after patiently enduring the steam heat under his net one man was found with his net flung back, quite contrary to orders. But all was forgiven when he explained that he had put down his net 'three nights running and never caught one yet'.

The weather was perfect and they led an idyllic life, except for the sandfly fever which struck a number down. But international matters were soon to occupy them, for in late July the Greeks, whose forces were then having some success against the Turks in Anatolia, threatened to march through Thrace on Constantinople itself with five divisions. The Allies hurriedly manned the formidable Chatalja Lines, a natural defensive position of great strength which ran from the Sea of Marmora to the Black Sea some 30 miles from the Bosphorus. Faced with the might of five British battalions, a force of French colonial infantry and the presence of the British Fleet, the Greeks thought better of it. Soon, as it happened, their armies were in full retreat, and the Greeks were then no longer a force to be reckoned with.

Of quite a different stamp were the Turks, and in early September they drove the Greeks back upon Smyrna (Izmir) and sacked the place. Now, with Kemal Ataturk in full cry, the tiny Allied force in Turkey was in grave danger.

Determined to keep open the Dardanelles, two battalions of British troops were sent to occupy a small perimeter at Chanak on the Asian side of the Narrows opposite Gallipoli. Here, with the might of the Navy and a small force of French and Italians – who later pulled out – the Allies waited for the advance of Kemal Ataturk's victorious armies.

In Constantinople, where there were known to be at least 20,000 armed Turks with more drifting in under the guise of peaceful travellers, trouble was expected hourly. The battalion were confined to barracks, while a couple of sections of Lewis guns stood by to intervene and prevent looting; but the Irish Guards' main task was to descend upon the police stations and disarm the police should it become necessary. In the end neither plan had to be put into action. There was a lot of glass broken by the jubilant Turks, but, happily, no concerted attempt to pillage the European quarter, where most of the shopkeepers were Greek.

Towards Anatolia a further defensive line was constructed east of Scutari on the Asian shore to protect Constantinople should the Turks attack from the direction of Ismid – but to the relief of the Irish Guards company defending the place they were not called upon. Urgent reinforcements had already been sent for, and by early October some four squadrons of assorted aircraft and a further seven battalions of infantry had arrived, including the 2nd Grenadiers and 3rd Coldstream. The three battalions of the Brigade of Guards then combined to form 1st Guards Brigade.

This vast reinforcement perhaps helped to pave the way for the Settlement at Mundania, which was signed at the end of October and which once and for all left control of the Dardanelles in Turkish hands.

With the crisis over, 1st Guards Brigade settled down to enjoy their stay in Constantinople. Sports of all sorts were the popular attraction: football, where they played in a cemetery, boxing, swimming and, above all, athletics. It was in the Brigade athletics meeting that, to the delight of the whole battalion, Alex won the mile race – a rare enough achievement for a commanding officer.

A flurry of excitement broke the calm when a little operation to evacuate the old Sultan was mounted. And while the 2nd Grenadiers were rather ostentatiously carrying out arms drill in their nearby barracks, and other men of 1st Guards Brigade casually walked around the area, two ambulances backed up to the rear entrance of the Sultan's Palace and Sultan Mehmed VI, Caliph of the Mussulmans, was spirited away.

Early in December occurred an historic occasion of a different sort, for the Navy gave a magnificent banquet on board HMS *Iron Duke* in commemoration of the first time since the Crimea that the Navy and the Household Brigade had served together.

Winter passed and the spring, and soon St Patrick's Day was upon them. Before the days of refrigeration a recurring dilemma was what to do with the shamrock in a hot climate. When three weeks before the great day a brown and withered consignment arrived and was shown to Alex, he was horrified. The Quartermaster, Captain H. Hickie, MBE, MC, rose to the occasion. 'Leave it to me, sir; I will get it right for the day.' But as the day approached there was still no sign of the shamrock and, curiously, whenever Colonel Alex raised the subject the Quartermaster was evasive. At length the Commanding Officer could bear it no longer. He must see it for himself. He and the Quartermaster wound their way down into the great cool cellars of the barracks, until finally they came on the shamrock, and there it was, beautiful and green and as fresh as though it had come out of Ireland that very morning. 'Hickie, you've achieved a miracle,' was Alex's comment, 'what did you do?' 'Well, sir,' was the reply, 'I just gave it some of what you and I like best. I gave it a sup of Irish whiskey.'

The battalion left Constantinople in October that year for Gibraltar, and remained there until the following April when they returned to England, and the unsalubrious Inkerman Barracks at Woking, which had once been a women's prison. A year later the Colonel of the Regiment, Field-Marshal the Earl of Ypres, died and was succeeded by General the Earl of Cavan.

The peacetime training season started with platoon and company training, rising through that for battalions and brigades and culminating in full-scale army manoeuvres in the early autumn. Motor transport was still almost non-existent, and in fact the only car or lorry used in the Irish Guards was driven by Mr Batts, the invaluable Officers' Mess caterer, who borrowed a private car from one of the officers for the period of field training and who, although a civilian, accompanied the Officers' Mess wherever it went.

They were starved of other equipment, too. It was true that the 1st Battalion Irish Guards possessed an anti-tank platoon, but they were mounted on bicycles and were armed with green flags! Another new feature was the Intelligence Platoon, some dozen men strong. They were drawn from the Pipes and Drums, were also mounted on bicycles, and still wore their kilts until the people of a Sussex village complained that their appearance in this dress on a bicycle was causing embarrassment; from then on they were ordered to wear trousers.

Tanks were simulated by cardboard shapes mounted on bicycles. Other groups of men were to be seen wandering about in canvas clothing bearing pole targets to represent sections, or machine-guns. There was little live ammunition for range-firing, and make-believe training was carried on relentlessly and seemingly without purpose. The Army was pilloried in books, in the Press, in plays and on the screen. It was a period of intense futility and frustration.

It was thus with some relief that the battalion heard in 1936 that they were to be posted abroad. In November of that year they sailed for Egypt for garrison duties under the command of Lieutenant-Colonel R. B. S. Reford, MC.

Before leaving England a number of men had been sent on drivers' courses, and as soon as they arrived in Egypt the battalion 'went mechanized'; while for their part the Intelligence Section found themselves issued with camels – a 'remarkably misconceived animal' was the verdict.

The Irish Guards had been in Egypt for close on eighteen months when early in July 1938, and quite unexpectedly, the Commanding Officer was sent for one morning: the battalion had been ordered to Palestine to help keep order between the Jews and the Arabs in what was known as the 'Triangle of Terror' in Central Palestine.

It was the first real taste of excitement the Micks had experienced since the Great War, and there were few of those veterans still serving. Everyone was eagerly awaiting a chance of meeting their enemy, the local brigands who were known as the 'ouzlebarts', or 'ouzles' for short. They weren't to be disappointed.

After a tedious and uncomfortable journey they moved into their camp near Nablus and on the third night they were there the 'ouzles' had the impertinence to attack. This was too much for the Micks, who let loose with every weapon they could lay their hands on. It was a grand battle; the fire was undirected and completely indiscriminate. The Brigadier was hopping mad, not least because his own camp was in the direct line of fire, and ordered the Commanding Officer to get things under control. Back came the reply: 'Oh, give them another five minutes, sir.'

With that out of their system the battalion settled down to a highly exhilarating little war against an energetic and skilful enemy.

It was an exciting four months for the battalion, now commanded by Lieutenant-Colonel W. D. Faulkner, MC, but the Munich Agreement put paid to their activities in Palestine and Egypt. The Irish Guards were recalled to England in November 1938. In the Palestine campaign they were awarded three Military Crosses, four Military Medals, and twenty-six Mentions in Dispatches.

They returned to an England arming for a war which semed inevitable. To give credence to this, on July 15th, 1939, the 2nd Battalion Irish Guards were reactivated and command given to Lieutenant-Colonel J. C. Haydon.

So great was the influx of Reservists and recruits to the Regiment that somewhere had to be found to accommodate and train them. This task was entrusted to Lieutenant-Colonel The Viscount Gough, MC, who had commanded the 1st Battalion in the early 1930s, and he managed to find out that Hobbs Barracks at Lingfield was then empty, so that became the home of the Irish Guards Training Battalion for the duration of the war.

For three years Lord Gough presided over the Training Battalion, and those who served under him hold a soft spot in their hearts for a really great 'character'. One day a man was arraigned before his Commanding Officer on a charge of drunkenness. 'Now, my man, just how drunk were you?' asked Lord Gough. To which came the shattering reply, 'Well, my Lord, begging your pardon, my Lord. I was as drunk as a Lord, my Lord!' The case was dismissed.

From the outbreak of war, and as the 'phoney war' continued, so the two battalions of Irish Guards gradually acquired their wartime shape. They knew they must be called some time, and speculation was rife as to where and when; but in the event it came sooner and in a totally unexpected theatre.

Chapter 8

The Phoney War

1 Norway

The German invasion of Norway began on April 9th, 1940. When the Norwegians awoke that morning it was to find that Oslo, Bergen, Trondheim and Narvik were firmly in the hands of the invaders or under heavy attack. But the Norwegian Government, despite the unhappy precedents of Poland and Denmark, declared that 'resistance would continue as far as possible'. It was a brave but hopeless gesture.

The two battalions of Irish Guards were then both stationed in Wellington Barracks and on April 1st the Commanding Officer of the 1st Battalion, Colonel Faulkner, received orders that his battalion were likely to move in the near future. On April 4th the King inspected them, thus giving credence to the rumours that they really were going. On the 8th the Bren-gun carriers were loaded on to flats at King's Cross station and vanished northwards. There was still no definite news of their destination, but April 10th was now being mentioned as the probable date of departure.

On the evening of that day the battalion paraded on the square in their new battle-dress and piled on board Green Line buses scrawled with such messages as 'To Norway', 'See the Midnight Sun', 'The North Pole Express'. So Norway it was to be. To the well-used valediction of 'Keep your head down, Mick' from their comrades of the 2nd Battalion, the 1st Battalion Irish Guards were piped from barracks.

Their train journey was memorable for the crates of beer

or double magnums of champagne in every carriage – the Micks were going to war in the heroic manner. This happy taste was to turn sour in their mouths before long as the dockers at Greenock stole all their binoculars and other valuables before they boarded ship. On the afternoon of April 11th the *Monarch of Bermuda* sailed under the battalion flag – hoisted under protest. In another ship of the convoy were the 1st Battalion the South Wales Borderers, also in 24th Guards Brigade; the remaining battalion, the 1st Battalion Scots Guards, had gone on ahead. The Brigade was commanded by Brigadier the Hon W. Fraser, DSO, MC, a Grenadier.

The voyage passed peacefully enough. Arctic clothing was issued on a lavish scale, and, as testimony to the powers-that-be, rarely were the battalion cold during their short stay in Norway; nor should they have been with eight extra pairs of socks per man, three pairs of gloves, two sweaters, two leather jerkins, a sheepskin coat which weighed over fifty pounds and a white furry hat, which was promptly christened a 'baby bearskin'.

The Commanding Officer summoned the officers to his cabin when they were well at sea. It transpired that while the true situation at Narvik was being resolved the battalion would land at Harstad forty miles to the north-west.

On the fourth day out they rounded the northernmost point of the Lofoten Islands and sailed down Vaags Fjord. From time to time appeared little fishing villages, with their wooden houses and insecure-looking jetties. And each miniature port had its own collection of 'puffers', the small fishing boats with primitive engines and independent-minded skippers which were to be of such invaluable service in the forthcoming campaign.

They reached Harstad at last and settled down to a life of ease and leisure in the local school. Here they heard the Germans announce over the captured Oslo radio that a body of hired volunteers and unemployed had landed in the north – this riled them; they preferred to be known as mercenaries.

Other reinforcements were on the way to help the hard-pressed Norwegians, and on April 20th a British brigade was landed at Namsos, 300 miles south of Narvik, under the redoubtable Major-General Sir Adrian Carton de Wiart, VC, and another at Andalsnes 150 miles further south. But they were of little use, for weakly armed and without supporting weapons, they found that bravery was not enough against a skilful and determined enemy who had unhindered use of the air.

In Narvik itself the German General Deitl, commanding the 3rd Mountain Division, which had recently fought in Poland, was firmly in control of the situation. His original force of some 2,000 men had been augmented by a similar number of sailors – survivors of the two highly successful British naval attacks on the German destroyers. With this force and the promise of whatever air cover he needed, General Deitl awaited the next Allied move.

At Harstad itself the rear echelons of the expeditionary force, the clerks, signallers, transport, and, of course, the staff, accumulated over the next few days. When the chaos in the harbour was at its height the German bombers came in to join the party. The Commanding Officer was holding a conference at the time. The officers emerged hurriedly to see the town in smoking ruins, for the wooden houses mounted on stilts to keep them clear of the snow collapsed like matchwood under the bombing.

The bombers came back the next day, and every morning for the next week, but by then the battalion were well away from Harstad.

The Irish Guards, travelling on HMS *Vindictive*, a converted repair ship, were ordered to Liland, a small fishing village in Bogen Bay north-west of Narvik and on the other side of Ofotfjord. There they waited for the orders to attack Narvik itself.

Number 2 Company were detached to occupy a little town, some five miles from the main battalion position. They set off with thirty-eight sleighs and after a number of adventures arrived in the early afternoon. By six o'clock in the evening, after Trojan work on the part of the signal platoon, Battalion Headquarters established heliograph communications with Number 2 Company. The first message was passed and took twenty-five minutes to decipher; it read: 'Captain D.-Matthews says will you return his towel immediately. You will know it by the name *Monarch of Bermuda*.' The Micks had arrived.

To pass the time they tried ski-ing, under the supervision of some officers and a Guardsman who turned out to be the reigning junior champion of Montreal. Some grew quite proficient – but most were confined to the toboggan slopes. There were other activities too, and the 'Big Irelanders', as they were known, made many friends.

On April 22nd the battalion were invited to attack Narvik. The plan had long been contemplated, promoted by the increasingly fiery signals from Winston Churchill at the Admiralty. But there was considerable divergence of opinion as to the best method of handling the operation, the first amphibious landing attempted by British troops since Gallipoli twenty-five years before.

In a howling gale, with snow and frozen spray whipping into their faces, the Irish Guards embarked in *Vindictive* the next morning. Almost as soon as they had left, a message arrived from Headquarters cancelling the operation, but there were no communications from the shore to the ship. Despite being signalled with torches, matches, anything that might attract attention, *Vindictive* sailed on.

Fortunately, the weather turned fouler still, and as the Navy felt themselves unable to guarantee the safety of the one landing-craft they had available, and with which it had been intended to land the Irish Guards one platoon at a time in what would have been an opposed landing, the operation was cancelled, to everyone's relief.

It was subsequently discovered that Narvik pier, where the first boat-load was to have landed, was heavily mined. After this fiasco the battalion returned to Liland once more.

It was a curious, indeterminate period. There were

fatigues to be done unloading the quantity of stores from the puffers at the pier, or attempting to create some form of bomb-proof shelter. But for the most part the battalion spent long, lazy afternoons fishing or boating or even bathing. The bombers came over regularly but they were used to them by this time.

The situation in Central Norway was collapsing. Driven from defensive position to defensive position by superior forces, the gallant survivors of the troops landed on April 17th were evacuated. On May 3rd a local capitulation was signed and all resistance south of Trondheim ceased. In a brilliant campaign the Germans had occupied the most important part of the country in a little over three weeks.

The German momentum barely slowed. Namsos itself, 100 miles north of Trondheim, fell on May 4th. By the 9th they were threatening Mosjoen, a further 100 miles northward. On May 10th it was learnt that the Germans had invaded France and the Low Countries.

In a vain attempt to produce some stability to a distressingly fluid situation 24th Guards Brigade were ordered south. The 1st Battalion Scots Guards were sent ahead and, after landing at Mo, advanced south towards the Germans. The Irish Guards received their orders to move on the 13th, and the battalion started embarking that afternoon on the *Chobry*, an elderly Polish liner. It was a tedious business as no arrangements had been made for them, so it was not until three o'clock the following morning that the last man and box was jammed aboard. And there they remained for the whole of the next day, a sitting target.

Eventually, at half past six, *Chobry* weighed anchor and proceeded slowly down the fjord. Their escort, the destroyer *Wolverine* and the sloop *Stork*, had preceded them and were already out of sight. It was a chilly evening, but the frustrations of the past days were forgotten among the comforts of the *Chobry*. 'This is the way to go to war,' said one officer. 'But it's all very well,' replied Colonel Faulkner, 'it only needs one bomb. It would go through this ship like a hot knife through butter . . .'

At midnight three Heinkels bombed the ship. The entries in the War Diary were laconic: '0015 15th May, HMT *Chobry* bombed and set on fire. Ship abandoned. 0903, arrival of *Stork* and *Wolverine* at Harstad.'

'The cabins collapsed like a pack of cards,' said one of the brigade staff officers who was aboard. 'Lights went out, the whole of the top decks amidships were immediately ablaze and very soon the main staircase seemed to have disappeared.'

The fire spread rapidly, for the explosion had wrecked the sprinkler system. The flames reached some ammunition stacked on deck, so lines were formed and they began chucking mortar bombs and ammunition cases overboard. The crackle of .303 rounds and the exploding flares of red and green Very lights loosed off to simulate anti-aircraft fire mingled with the steady firing of the A/A Brens.

After the first explosions the battalion were ordered on deck. The fire had divided the ship into two halves, with the result that most of the men had to go forward and all the officers, except Father Cavanagh, aft. The companies filed up from their mess-decks, in full kit, carrying their rifles and Brens. 'Get on parade – face that way.' At the sound of Regimental Sergeant-Major Stack's familiar voice the men formed up in mass, Drill-Sergeant Hoey in front, Drill-Sergeant Peilow in rear. The sentries already on deck acted as markers. Only one man broke rank and jumped overboard and he broke his neck on his life-jacket.

The lifeboats forward could not be lowered, for the power was cut off and the electric winches were unworkable; even the herculean efforts of the Guardsmen on hand winches proved of no avail. There was nothing to do but wait till the escorting destroyer came alongside. Father Cavanagh had been told by a naval officer that whatever happened the men must not be allowed to jump into the icy water, which would have been certain death unless they were picked up in a matter of minutes. Realizing that somehow he had to keep the men occupied, the priest started to recite the Rosary, and with bared heads on a burning ship in the

Arctic Circle men said the prayers that they had learnt long ago in the quiet churches and farmhouses of Ireland. It was something familiar and spiritually comforting and they forgot, momentarily, the great peril they were in.

By the time the escort came up there were some twenty heads bobbing about in the sea, for the fire amidships was spreading and forcing isolated men overboard. The *Stork* launched her whaler to pick them up. She herself stood off astern, engaging the bombers with her guns, and taking on board the men from the lifeboats, while HMS *Wolverine* came alongside the stricken *Chobry*.

'We closed on their burning and sinking ship,' said Commander Craske, RN, Captain of the *Wolverine*. 'I never before realized what the discipline of the Guards was. We got a gangway shipped forward and the men were ordered to file off on to us. There was no confusion, no hurry, and no sign of haste or flurry. I knew that there might be only a matter of minutes in which to get them off. I had four ropes fixed so as to hurry up the transfer. They continued to file steadily off in one line. I cursed and swore at them, but they had orders to file, and they filed. I saw someone who seemed to me to be a young officer and in no measured terms I told him to get them off by all four ropes. In a second they conformed to this order by one of their own officers, still steadily and without fuss or confusion. Their conduct in the most trying circumstances, in the absence of senior officers, on a burning and sinking ship, open at any moment to a new attack, was as fine as, or finer than, the conduct of the old days of the soldiers on the *Birkenhead*. It may interest you to know that 694 men were got on board in sixteen minutes.' Some still possessed their rifles, for no mere naval disaster could shake the training of the Guards Depot, backed by a sergeant shouting at them, 'Take care of those rifles, they are Government property.'

Two years later Admiral Craske wrote to the Regimental Lieutenant-Colonel: 'You will not know me. My son, Commander R. H. Craske, RN, was in command of HMS *Wolverine* in May, 1940. He lost his life in HMS *Barham* in

November, 1941, and among the papers he had treasured is your letter to him referring to the rescue of the battalion from the *Chobry*. He never mentioned the affair till the following December, when he was awarded the DSC, and they told us very little, except that any rescue would have been impossible but for the superb discipline of the men of the battalion.'

The 'letter' was a message from the Officer Commanding, 1st Irish Guards, dated May 16th, 1940: 'I wish to convey to you the deepest gratitude and admiration of all ranks for the bravery and skill displayed by officers and crews of HMS *Wolverine* and *Stork*. It was entirely due to your gallant action that so many lives were saved.'

'The sailors were wonderful,' wrote an officer, 'and made cocoa and gave us clothes. We had a seven-hour journey back to Harstad and were bombed all the way. Not a good ending.'

The *Wolverine* arrived at Harstad first and discharged its load of survivors on the quay. They came ashore wrapped in blankets and naval greatcoats over pyjamas and dungarees. Stretcher-bearers carried the wounded off first, and laid them in a little tin shed. From there the worst cases were taken straight to the Base Hospital. Drill-Sergeant Peilow jumped ashore and doubled away to look for a square. He found one and put out company markers, while the RSM ordered the men straight on parade. They fell in – 'a proper battalion parade' – and waited for the other ship to arrive. The Scots Guards, still at Harstad, with real charity sent down containers of tea and mounds of 'wads'. The officers and the Company Sergeant-Majors waited on the quayside for the *Stork* to come in. Everybody was desperately hoping that the Commanding Officer would be on board. As the *Stork* came in, they saw Captain Gordon-Watson in pyjamas standing on the bridge and he shouted inquiries. 'Where's Colonel Faulks?' 'No sign.' 'Jack Durham-Matthews?' 'Never seen,' and so on. 'I almost cried,' he wrote.

At one blow the 1st Battalion Irish Guards had lost their Commanding Officer, Second in-Command, three Company Commanders and the Adjutant, killed; the other two Company Commanders were wounded and out of action for the rest of the campaign. As senior officer present, Captain Gordon-Watson took over command. He wrote to Regimental Headquarters, 'We have lost all our equipment, except what we stand up in. The drums, fifes and bugles were not in the ship and we have saved five pipes...' He ended his letter on the most pathetic note of all: 'Could you send us out a nominal roll of the battalion?'

It was a tragic loss, and the battalion felt only an awful emptiness and overwhelming sense of desolation. For the rapport between officers and their men is perhaps closer in an Irish Battalion than in other regiments. Their loss was thus one of personal friends more than of superiors.

But their greatest loss was Colonel Faulkner, known as 'Faulks' throughout the Regiment, who was revered by every man in the battalion. He was an enlightened Commanding Officer and his death came as a terrible blow. History had repeated itself in bitter fashion: the 1st Battalion Irish Guards had again lost a deeply respected Commanding Officer at the outset of a major war.

The other battalion in the brigade, the 1st South Wales Borderers, also had their troubles, for the cruiser *Effingham*, on which they were embarked, struck a rock in the Leads and sank, fortunately with little loss of life.

On May 20th, in the middle of a bombing raid, the Irish Guards left Harstad for a second time, this time by a mixture of destroyers and puffers which had been the original plan advocated with some force by Colonel Faulkner. They reached their destination near Bodø around midnight, and after a fruitless search for Brigade Headquarters settled down for what remained of the night among a miscellaneous pile of equipment salvaged from the *Effingham*. The Quartermaster and his henchmen spent the next hours making up deficiencies, so that by the morning the battalion were more or less fully equipped for battle.

Dawn and a stray staff officer arrived together; the latter supposed that the local headquarters might be some ten miles farther down the fjord. There they duly found the remainder of the British troops in the area all under the command of Colonel C. McV. Gubbins, MC, who had delegated local authority to Lieutenant-Colonel H. C. Stockwell.

The position to the south was becoming increasingly precarious. It would have been difficult enough with a full brigade to hold the Germans, who were by now over 1,500 strong and who had complete mastery of the air. But the disasters to *Chobry* and *Effingham* meant that only those troops who were already in the area could stem the enemy advance.

Fighting a series of gallant delaying actions, the Scots Guards worked their way back northwards under round-the-clock air attacks. And there was little sleep in the Arctic twilight when the enemy could keep their every movement under constant observation.

Aware that it was but a question of time before the Scots Guards, spent with the constant depressing business of a retreat against overwhelming forces and in the face of the unrestricted enemy air effort, would soon no longer be able to hold up the enemy, it was decided that the Irish Guards would man the next defensive position.

Pothus stands on the Saltelva river some ten miles from Rognan, the jumping-off point for the ferry to Bodø. The river in spring, swollen by the melting snows, is an unfordable torrent, and in 1940 it was spanned by an iron-girder bridge which carried the main road from the south. The bridge was dominated by a razor-backed and thickly wooded ridge through which the road continues until it reaches Rognan. It was here that the Commanding Officer, Captain H. C. McGildowny, who had rejoined the battalion at Harstad, decided to make his stand.

He had under command the four companies of the Irish Guards, a troop of 25-pounders and a section each of Norwegian mortar and machine-gun units. There were three Independent Companies, of British Territorial volunteers, posted to picket the heights in the hope of at least hindering the Germans in their outflanking tactic. But none of the troops present had seen or experienced the hampering effect of unrestricted air-power, for the whole position was in comfortable range of the advanced German fighter bases.

It was decided to place Number 1 Company across the

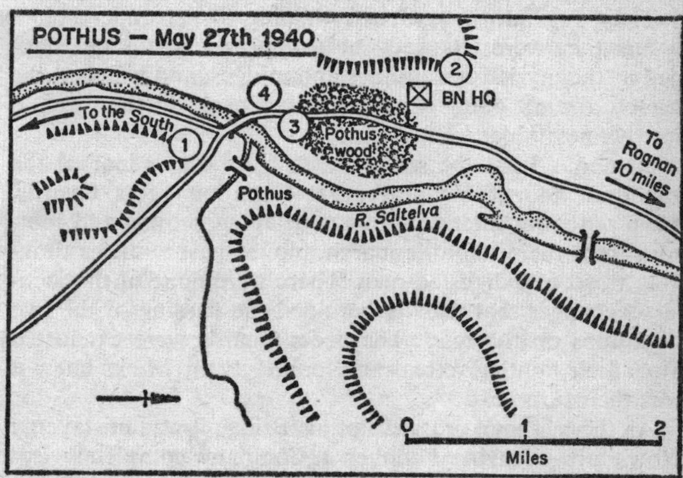

river to look after the approaches to the bridge on the eastern side of the river. The remainder of the battalion and the Norwegian support were put in and around Pothus Wood. Here they could watch the bridge and the other most likely line of advance across the hills.

It was a magnificent and typical spring morning, when Norway is at her most beautiful best, and there was little sign of war as the Irish Guards trudged up the lovely Saltelva Valley. Only the miniature white farmhouses with their walls riddled with bullet holes betrayed the fact that the Germans were not far away. That they were active again that morning became clear as the leading company reached the outskirts of Pothus; from then on progress was through the woods and in the ditches.

The position of Number 1 Company across the river was

undesirable to say the least. Out of reach of small-arms support from the remainder of the battalion, they were also virtually out of touch, for the primitive wireless didn't work, and the telephone cable extended only halfway to their position. Communication, such as it was, was by runner.

They dug into the soft soil, hoping to get underground before they were attacked, sat around in the sun and slapped at the myriad mosquitoes and midges, and watched the enemy aircraft going about their business. So they waited for the inevitable.

At about 1 AM the sentry nearest the road reported the sound of footsteps and the 1st Battalion Scots Guards, worn out from their continuous delaying actions and footsore from their 100-mile march, moved past with the dragging tread of exhausted men. There were none of the jocular exchanges that usually heralded the passing of the two battalions on the road. The Scots Guards were exhausted after their fighting retreat and looked it; the Micks knew it was their turn next.

As the last man crossed Pothus Bridge it was mistakenly blown up – leaving Number 1 Company completely isolated from the rest of the battalion. An uneasy silence settled on Pothus.

The morning wore on and as the sun came up the mosquitoes came out in force. By 8 AM it was clear that there was a considerable body of enemy in front of Number 1 Company. The enemy mortars were now ranging on the defenders, and small-arms fire was making life uncomfortably hot in the forward platoons.

Early in the morning the first Heinkels appeared and flew suspiciously close to Battalion Headquarters sheltering in Pothus Wood. It was a trial run. Sweeping low they then proceeded to machine-gun the wood. 'Up and down they went, raking it with the methodical accuracy of ploughmen.' Battalion Headquarters, by the road on the eastern edge of the wood, watched this fascinating performance. They lay on the ground in a long line. 'When I blow the whistle,' shouted Regimental Sergeant-Major Stack, 'jump up and double across to the far side.' While the Heinkels were climbing to turn, the long line of men raced through the wood and dropped panting behind the bullet-scarred trees. One Guardsman was slightly wounded. Meanwhile a number of dive-bombers had also made their unwelcome appearance, but the two-man slit-trenches which they had learned from their Norwegian hosts to dig saved them any casualties. (For his fine leadership throughout the Norwegian campaign and particularly for the part he played in the rescue of the battalion from the *Chobry*, Regimental Sergeant-Major Stack was awarded the Military Cross; only the second warrant officer to gain this award in the history of the Regiment.)

As this activity was going on, the Germans mounted another and heavier assault against Number 1 Company. It was driven off. But it was clear that it was only a question of time before an overwhelming attack was thrown against them.

In the later afternoon Captain B. O. P. Eugster, MC, the Company Commander, ordered one of his platoons to withdraw across the river. Half an hour later he decided to withdraw the remaining platoon. Everywhere the river, swollen by the melting snow, was running fast and deep. The Germans soon discovered that the position had been abandoned. They came up over the ridge and opened fire on the platoon down below them. The Guardsmen whipped the slings off their rifles and knotted them into a chain. One tied the end of the chain around his waist and plunged into the river. The racing current swept him a hundred yards downstream before he struck a small island near the far bank. Captain Eugster anchored himself in the middle of the current to help each man across, while Platoon Sergeant-Major Thompson remained on the bank and threw into the water any man who hesitated. All the time the Germans were getting closer and their rifle fire more accurate. When the whole platoon was across and away, Platoon Sergeant-Major Thompson joined Captain Eugster, who had been wounded in the shoulder at some stage in the battle, on an island in the middle of the river. There the two of them lay, shivering miserably for three hours before the fire slackened. On the way back to Rognan they found a boat from somewhere, and crossed. Captain Eugster, swinging his revolver in his good hand, stalked into Battalion Headquarters to report, and was promptly evacuated to the hospital at Bodø.

With their usual efficiency the German mounted troops exploited Number 1 Company's departure. They managed to cross the river and were soon engaging the main battalion position itself. The situation was most unhealthy. By midday on May 27th it was clear that the Irish Guards were in danger of complete encirclement. So Colonel Stockwell gave orders for a general withdrawal.

'At midnight,' says the War Diary, 'the battalion, less those unaccounted for, left Rognan by puffer for Fineid.' Those unaccounted for included the whole of Number 2 Company – who had been ordered to hold a hill away from the main battalion position and had never received word to withdraw – half of Battalion Headquarters and a few men of Number 3 Company. These remnants began to assemble in twos and threes at Rognan. Soon only a small party of Battalion Headquarters were missing – and they turned up later. The battalion's overall losses had been astonishingly small.

The Irish Guards occupied a temporary position on the Valnes peninsula with the rest of Colonel Stockwell's little force. There they heard of the capture of Narvik by a mixed brigade of Norwegians, Poles and French. But on their own front it was clear that they themselves must pull back, and a few days later they returned by puffer to Bodø.

At midnight on the day of their arrival a message came up from Bodø harbour, 'Destroyers *Firedrake* and *Fame* at quay. Five minutes to get aboard.' Sailors standing at the head of the gangway ruthlessly threw overboard anything they considered superfluous. They seized the large tea-chest full of the battalion documents, giving the orderly room clerks just time to snatch out a nominal roll and a casualty list before throwing it into the sea.

Resistance in France and Belgium was collapsing, and it was clear that soon all available fighting men would be needed to defend Britain herself. So, reluctantly, on the evening of June 1st the decision to evacuate the Expeditionary Force was communicated to King Haakon and the Norwegian Government. On the 7th the King embarked from Tromsø on the cruiser *Devonshire* and on the 8th the evacuation began from Harstad.

For the last time the Irish Guards were ferried from shore to ship by the faithful puffers, and in the *Lancastria* the battalion crossed the North Sea to make their landfall at Greenock two months to the day from when they had left Wellington Barracks.

They returned with stories galore, only to discover that in their absence the 2nd Battalion had also been committed and had been involved in even fiercer battles than they.

II The Hook of Holland

The departure of the 1st Battalion for Norway had hardly ruffled the tenor of existence of the 2nd Battalion in Wellington Barracks. While they performed Public Duties or a little gentle field training in St James's and Hyde Parks – in the full gaze of a fascinated audience – they wondered when their call would come.

Throughout Britain battalions were being hurriedly assembled to form brigades and divisions to resist a possible invasion. So it was to the 20th Guards Brigade, then being formed at Old Dean Camp near Camberley, that the 2nd Battalion Irish Guards were sent to join the 2nd Welsh Guards and the 5th Battalion the Loyal Regiment, the whole under command of Brigadier W. A. F. L. Fox-Pitt, MC, himself a Welsh Guardsman.

Whitsun leave arrangements occupied their attention – but no sooner were the lucky ones away than an urgent message arrived from the War Office cancelling all leave, for that very dawn the Germans had invaded the Low

Countries and it was clear that every man would soon be needed.

But training had to go on and there was no thought of cancelling the night exercise on Friday, May 10th. The battalion marched back the following morning in time for a late breakfast. After they had eaten and while they were still cleaning their weapons and equipment before settling down to rest, the order went out: 'Pack up, pack up, prepare to move.' The purpose of this unexpected order was to get the brigade concentrated in an area nearer the South Coast where it was to be employed on home defence duties or, more pertinently, anti-invasion activities.

It was not until nearly midnight that the last man of the battalion had found his sleeping quarters in the oast-houses and outbuildings of a farm near Paddock Wood in Kent. But there was no sleep for the Commanding Officer that night. Summoned to Brigade Headquarters, Colonel Haydon was given brief orders. 'You will command a composite battalion of this brigade. This battalion will embark at Dover for Holland tomorrow afternoon. Report to the War Office in the morning at ten o'clock for further orders.'

The subsequent orders were a little more explicit, and Colonel Haydon was told that a force of 200 Royal Marines had landed at the Hook of Holland during the night with orders to secure a landing place. On arrival he himself was to get in touch with the local Dutch commander and then to move towards The Hague with the object of safeguarding the Dutch Government and restoring the situation in the capital.

The battalion embarked at Dover on Sunday evening, May 12th. The Adjutant had managed to contact the authorities at Fishguard and Holyhead and some of the leave men were sent back before they reached Ireland. But they were in civilian clothes and their uniforms were at Old Dean Camp, so a number of Micks went to war wearing neat, highly polished pointed civilian shoes under a battledress which they managed to scrounge from somewhere. Even so the battalion were still well below strength and a 'company' about 200 strong commanded by Captain C. Heber-Percy was borrowed from the 2nd Battalion Welsh Guards.

At the last moment a large packing case arrived containing maps and a special wireless set for direct communication with the War Office, and with it a Dutch interpreter.

Harpoon Force, which was now their official code-name, spent a miserable night on a rolling sea and it was at dawn on Whit Sunday, May 13th, 1940, that the battalion sailed into the harbour of The Hook of Holland, wondering what sort of reception they would receive and what was in store for them in this the first action of the reformed 2nd Battalion.

The Commander of the Royal Marine contingent met them, but the only news he could give was that the position was obscure. Rotterdam, only a dozen or so miles away to the east, was clearly under heavy attack, for great columns of black smoke could be seen and there was a glow in the sky from the already burning city. It was not a happy augury.

Shortly after Colonel Haydon took up residence in the Hotel Amerika in the main square, an urgent call came from the British Military Attaché at The Hague: 'Could the battalion please stay in The Hook, guard the port and keep open the road to The Hague?' The reason soon became clear.

It was now about 11.30 on the morning of May 13th and an event of historic moment was about to unfold.

Suddenly, a fleet of large black cars escorted by a number of motor-cycle outriders swept down the road and drew up sharply by the quay. Out of one of the cars stepped the unmistakable figure of Queen Wilhelmina of the Netherlands. While the Commanding Officer stayed to receive the Queen, the Adjutant was dispatched to warn the Senior Naval Officer – Captain W. F. Halsey of HMS *Malcolm*.

Captain Halsey returned with the Adjutant to the quayside and it soon became clear that the Queen – resolute and indomitable in the grave crisis facing her country and herself – was desperately anxious not to quit Dutch soil but, instead, wished to be taken to either Flushing or one of the Dutch islands close to the mainland. Captain Halsey's desire was, of course, to do all he could to give assistance, but he had to point out that though he was aware of the location of the British minefields, he had no knowledge of where the Dutch ones might be. This being so, he could not possibly undertake the risk of taking the Queen to Flushing or to one of the islands unless a chart of the Dutch minefields could be made available. None of the Queen's party had such a chart and so, in the outcome, there was no alternative but for that brave and gallant lady to take passage in a British destroyer to an English port.

The departure of the Queen removed one of the reasons for the battalion's presence in Holland. The arrival of the Dutch Government, closely followed by the Diplomatic Corps, removed another, and a message at last came through from the War Office to the effect that steps were being taken to evacuate the small force.

The morning of May 14th brought only the Germans. Mounting their heaviest air attacks so far they systematically flew up and down the straight little streets of the village, machine-gunning mercilessly. The refugees were the target. Father Julian Stonor, the Battalion Chaplain, takes up the story: 'The rest of the morning [May 14th] I can only remember as a confused kaleidoscope of horrors. I found myself facing an endless stream of refugees from poor Rotterdam. Some of them were pushing dead relatives or children on handcarts and perambulators, and every now and again a German plane would swoop down and machine-gun some more. But there were no signs of panic; in fact no one even bothered to take cover in the ditches beside the road. Many of them were Catholics, and I was able to give them Absolution.' He himself, as unmoved as were the refugees, was a strange sight as the crown of his steel helmet had somehow come off, leaving him with merely the brim settled like a collapsed halo.

Conscious that there was no salvation at The Hook,

some of the fugitives had already turned north towards The Hague, running the gauntlet of the German aircraft for a second time.

Time, too, was running out for the Irish Guards. It was abundantly clear that to prolong their occupation of The Hook would do no good, and was risking the whole force. Already there were rumours that the Germans had occupied The Hague and were heading southwards in their direction. An armoured thrust would have been fatal, as they had no anti-tank guns with them. Thus, when the Senior Naval Officer announced that ships had become available for the evacuation, Colonel Haydon ordered the operation to begin forthwith.

At ten o'clock that night they landed at Dover and the following day entrained for Camberley. They drove into their old camp through lines of cheering Welsh Guardsmen. 'An embarrassing experience,' remarked Father Stonor, 'for we had been able to do nothing for the poor Dutch.'

Eleven Irish Guardsmen are buried in the little village of The Hook of Holland. As for those who came back in the sweltering sun of an English heat-wave they found it hard to believe that it had not all been a nightmare; six days later it occurred again.

III Boulogne

Holland had fallen and it was now the turn of Belgium and France. On May 17th Brussels was captured, and by the evening of the 18th the British Expeditionary Force was already threatened by the growing salient in the Allied line to the south, towards Amiens. On the 20th German tanks reached the sea near Abbeville.

At this crucial moment, and to cover the withdrawal and evacuation of the BEF from the Channel ports, reinforcements were hurried from England. A mixed brigade of three battalions of the Greenjackets were rushed to Calais, and 20th Guards Brigade were ordered to stand by.

The night of May 22nd had once again seen that brigade involved in a night exercise. No sooner had they returned than again the cry went up: 'Pack up. We're on the move.' They thought it was a joke, but it wasn't and at 3.30 PM they were away – but this time comprising complete battalions of the 2nd Irish Guards and the 2nd Welsh Guards, plus Brigade Headquarters.

The embarkation at Dover that night went smoothly enough, after the Commanding Officer had managed to convince the authorities that a ship designed to take 500 men could not take 700. But they were without maps, wirelesses, mortars or grenades, and with only four two-pounder anti-tank guns with which to take on the might of the German armour which they were told was advancing on Boulogne – their destination.

Arriving at Boulogne with only two battalions available and no more troops forthcoming, it was only a question of which half of the long perimeter to allot to which battalion. The town was conveniently divided by a conspicuous railway line; the area to the south and west of this mark was given to the Irish, while the Welsh had the formidable task of blocking the main access roads from the east and north. 2nd Battalion Irish Guards were ordered to assemble at Outreau, a suburb of Boulogne astride the principal approach from the south.

In single file and with all the ammunition they could carry, the companies forced their way up the hill against the tide of refugees flowing down towards the harbour. One company and the anti-tank platoon, which had travelled on a different ship from the rest of the battalion, had still not arrived, so the remainder were spread in a two-mile arc from the sea on the right to Outreau and the railway on the left. In these positions and benefiting from their brutal experience of German air-power at The Hook, they dug as they had never dug before.

The missing company duly arrived, a bare half-hour before the Germans, and they were still striving to get dug in when the first artillery bombardment started.

The enemy attacked in earnest later that afternoon, but by midnight the main positions still held, although casual-

Recruits in 1901

Number 8 Company's barrack room, Oudenarde Barracks, Aldershot, 1904

Recruiting poster, 1915

The trenches, Laventie, 1915. Capt the Hon H. R. L. G. Alexander. In background Lt R. Rankin and CSM Daly

Yser Canal, July 21st, 1917. Wounded Irish Guards Officers. 2/Lt G. E. C. Vaughan-Morgan (on stretcher), in attendance Capt D. Lees, DSO, MC, RAMC. Looking on, Pte Hannon, Irish Guards

Leitrim Boy receives his shamrock, St Patrick's Day, 1917, Warley Barracks. L to R: Drummer Titman, Master Vickers, Drum Major Fitzgerald

Intelligence Platoon, 1st Battalion Irish Guards, Egypt, 1938. L to R: Lt G. A. M. Vandeleur, L/Sgt Stone, L/Cpl Batt, Drummer Smith, L/Cpl Thompson, Drummer O'Brien, Guardsman Simpson, Drummer Thackrah

The magnificent feat of L/Cpl Kenneally at Bou Arada (Hill 212) which won him the Victoria Cross, April 30th, 1943

The Queen's Birthday Parade, June 10th, 1966. Capt B. J. Gilbart-Denham commanding the Escort to the Colour

Guard of Honour at the Coronation of Her Majesty Queen Elizabeth II, June 1953. Irish Guards Officers: Capt C. W. D. Harvey-Kelly, 2nd/Lt J. E. L. Nugent, Lt R. A. C. Plummer

Her Majesty the Queen at the Regimental Garden Party after the Presentation of New Colours, June 10th, 1966.
L to R: Sgt Matthews, Maj Carleton-Smith, RSM Sullivan, Miss L. Mercer, Mrs Sullivan, RSM Williams, Mrs Williams, Drummer Howard with Shaun

Her Majesty Queen Elizabeth the Queen Mother with Members of the Regimental Association at Windsor on St Patrick's Day, 1970.

Hong Kong, November 1971: in charge of captured guerillas

ties had been heavy and one complete platoon had been cut off together with the two precious anti-tank guns.

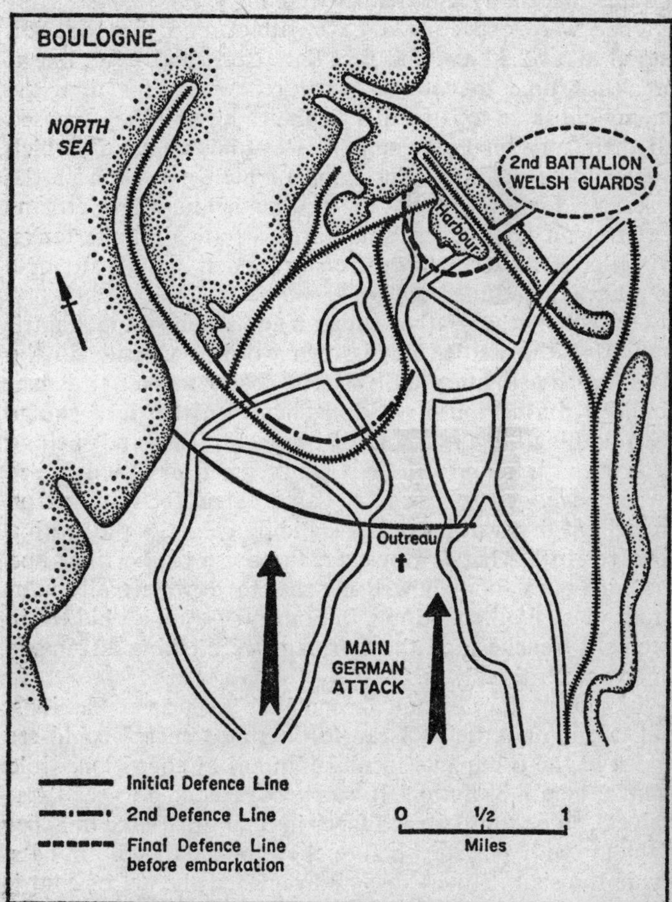

It was an uneasy night, for there was no information from Brigade Headquarters and none of the seven dispatch riders sent to find the Brigadier succeeded in doing so. Thus, with no wireless or telephone lines available, the battalion was in complete ignorance of what was happening, though judging from the heavy firing to their left they did assume that the Welsh Guards were busy.

They were expecting a dawn attack and the battalion 'stood to' at 2.30 AM, and first light came and went, but at breakfast time the attacks were renewed. This time the enemy chose an axis further to the right, and they focused all their fury on Lieutenant P. Reynolds's platoon which was holding an important knoll of high ground near the reservoir. Fighting grimly, the platoon held up the Germans for over an hour, using small-arms fire against the tanks. But it couldn't last; they were surrounded and all attempts to relieve them failed.

It was here that Father Julian made an individual contribution to the battle. In his own words: 'Outside in the daylight the situation didn't seem nearly so bad as it had sounded during those two long hours in the dark garage with nothing but the sound of incessant firing and shelling growing nearer and nearer. I think I had expected to see corpses everywhere, but to my amazement the only person in sight at first was an officer standing behind a wall with a revolver in his hand. However, the noise of the rifles and machine-guns soon showed me that there were really hundreds of men all round me, but our men were all hidden in deep slit trenches and the Germans were taking advantage of any cover there was.

'Obviously there would be no time to hear confessions, but taking my crucifix I ran to the first trench I could see and told the occupants to make an act of contrition while I gave them Absolution. It was very consoling to see those half-dozen or so up-turned faces in their grave-like trenches light up with pleasure before they took off their tin-hats, made the sign of the Cross and repeated the act of contrition they had learnt as children in Ireland. Any number of them, and not all Catholics, told me afterwards that from that moment onwards they felt no fear.

'With such a deafening noise of machine-guns I never thought I would be able to do more than one or two trenches, but as each trench directed me to the next and still nothing hit me I, too, lost all sense of fear; all the more exposed a man was, the more exhilaration I got out of going to see him.'

The effective strength of the battalion was now only two and a half companies, and the line was far too long to hold unaided. So gradually, and under heavy pressure, they pulled back to the light railway which at that point rings the town. Here they stood for two precious hours, and when the Bren-gun barrels became so hot that they warped and jammed they fought with rifles alone. But it was heart-breaking work trying to hold back the enemy who could infiltrate through the widely spaced sections positioned in the heavy cover which bordered the railway; and in response to orders from brigade, at about 1 PM the perimeter was pulled back still further towards the centre of the town.

Father Julian takes up the story: 'The word now went round that we were to be evacuated straight away, and as we marched in single file down the empty glass-strewn streets there was much speculation on our chances of finding ourselves back at Old Dean Common once more. The chances seemed very remote, and would have seemed more so had we known what was still in store for us.'

There was room for only two destroyers in the harbour at a time. While the mass of mixed troops of all regiments and nationalities embarked, 20th Guards Brigade, which consisted by this time of the equivalent of only two companies each of Irish and Welsh Guards, held off the enemy who were now right on top of the harbour. The destroyers *Keith* and *Vimy* came alongside first and, although under direct observed fire from the German batteries on the high ground behind Boulogne, the embarkation was continued.

When fully loaded, *Keith* and *Vimy* backed away. It was now the turn of *Whitshed* and *Vimiero*, and all was miraculously quiet as the destroyers glided toward the quay.

The two ships embarked all the troops they could and then cast off, each with some five hundred men on board, but few Guardsmen. However, some that did get aboard were the ten survivors of the anti-tank platoon which *Whitshed* had landed the previous day. A burly Guardsman climbed on to the bridge and deposited a bottle of champagne on the chart table. 'Thanks for the double ride,' he said and hurriedly disappeared. HMS *Whitshed* subsequently presented her bridge-emblem to the 2nd Battalion as a memento of this day and of their earlier journey with the battalion to The Hook and back.

Wild Swan and *Venomous* were the next destroyers to come alongside and the battalion moved down to the quay in sections and platoons. To speed the evacuation *Verity*, too, moved in. From seaward she had watched her sister ships pass safely into the inner harbour. All was quiet when she herself went ahead towards the breakwater. When she was within a few yards of the entrance there was a sudden sheet of flame from the low hills to the north as a hidden battery opened fire. This was followed by another, and a general bombardment broke out. The intention was obvious, for if the Germans could manage to block the harbour entrance evacuation would be impossible.

Verity now attracted all the enemy fire. Hit again and again she was soon an inferno. With great gallantry the surviving officer on her bridge ordered full-speed astern. Slowly *Verity* responded and backed away; the flames spread rapidly but she reached the open sea and so saved the troops at Boulogne from certain destruction or capture.

The only part of the port now in British hands was the quay itself. German tanks and infantry moved on to a parallel dock and commenced to pour a hail of bullets into the three destroyers only partly hidden by the solid masonry.

'It says a great deal for the discipline of the troops,' wrote the Commanding Officer afterwards, 'that no move of any sort was made towards the destroyers until I gave the order, and then the move was carried out slowly and efficiently.' A naval officer broadcast a description: 'The courage and the bearing of the Guardsmen were magnificent, even under a tornado of fire with casualties occurring every second. They were as steady as though on parade and stood like rocks, without giving a damn for anything.' The battalions in turn were filled with boundless admiration for the sailors manning their guns on the open decks. Their gunnery was splendidly effective. The *Wild Swan* sent a large German tank on the opposite quay spinning like a cartwheel, and blasted the top floor off an hotel when she spotted a machine-gun firing on the battalion from an upper window.

The wounded were slowly lowered from the quay and taken below, and then it was the turn of the survivors of the two battalions. At 9.30 PM the last of the Bren-gun teams doubled down the quay and leapt on board *Wild Swan*. The destroyer cast off and made for the open sea.

Described Father Julian: 'It was dark when I came on deck from the inferno below. I found a seat on a torpedo tube with some Guardsmen and we talked about the sea and our admiration for these sailors, until the flickering light that was the signal lamp of the wounded *Verity* was joined by another which proved to be the *Dover Castle*.'

It is not too much to say that for the second time in ten days the battalion owed their existence to the magnificent conduct and seamanship of the destroyers of the Dover Flotilla.

Five officers and 196 other ranks were killed, wounded or missing in the rubble and mess of Boulogne. The Welsh Guards, compelled to leave behind two complete companies who courageously fought on for another two days, lost about the same. But the two battalions were luckier than the Calais garrison; for there the battalions of the Rifle Brigade, 60th Rifles and Queen Victoria Rifles, were killed or captured to a man.

Chapter 9

Interlude: The years of preparation

The two service battalions of the Irish Guards returned to an England seething with home defence plans. It was a time of unparalleled and seemingly continuous disaster. On June 10th Italy entered the war determined not to miss the opportunity of gaining the spoils of another's victory; on June 22nd the defeatist elements in France sued for peace. Now only Britain remained, and the battle for Britain was about to begin.

The 1st and 2nd Battalions were in their home stations of Northwood and Woking respectively. They practised home defence schemes of varying complexity, and prepared themselves for what they and everyone else felt was imminent invasion.

Over the previous months Irish Guards reinforcements had poured in. These were in the Training Battalion at Hobbs Barracks, Lingfield. Here was a fully trained force, uncommitted to any brigade, and it was decided to send them to support the undermanned and half-trained troops which had been all that could be spared for coastal defence while the battle for France was at its height.

So, leaving behind a small detachment to look after new recruits, they moved to Dover. It was an inspired choice. The garrison duties there were not sufficiently onerous to interfere unduly with their training; they were able to experience a form of battle inoculation which was invaluable; and they had excitement of one sort or another all the time. They reached their new station in July 1940; and were to be there throughout the Battle of Britain.

Every day the Bren-gunners had a grandstand shoot at

enemy aircraft. The War Diary proudly records that 'Major Gordon-Watson shot down a Messerschmitt 109 on August 31st, and the guns of Number 3 Company a Dornier 17 bomber on September 29th.' There were also a number of unconfirmed 'probables'. 'On how few occasions,' the diary continues, 'can young soldiers have been provided day after day with such targets as low-flying aeroplanes manned by real Germans intent on the conquest of Britain? The roar of the barrage was a fine novelty for the young Irish Guardsmen preparing to join a service battalion.'

The Battle of Britain had been won by the end of October. A few weeks later, when the 'invasion weather' had broken, the Training Battalion, somewhat reluctantly, left Dover. They were bound for Northwood and there under the initial guise of 'The Holding Companies' the 3rd Battalion Irish Guards came into being. The first Commanding Officer was Major Vandeleur, and he was to remain in command of the battalion until late in 1944.

The next milestone in the history of the Regiment and in that of the Household Brigade was the formation of the Guards Armoured Division. The speed and skill of the German conquest of France, spearheaded by superb armoured formations, had impressed everyone. The call for armoured forces for the campaign then being waged in North Africa had left the home defence commands woefully short of armour. It was essential to convert at least two infantry divisions to an armoured role; what more natural than to create a Guards Armoured Division?

Although infantry by tradition, the existing organization of the Brigade had two unique advantages; in the first place there were a number of independent Guards Brigades. Unattached to any division they could thus be easily converted with the minimum of disruption to existing formations; and secondly, every man would have received a common grounding at the Guards Depot – and with such a starting-point armoured work would be merely a form of development training.

The idea was warmly received; the concept as well as

being revolutionary for the Brigade was obviously the rising fashion. It was a tremendous challenge and they were determined to prove to the denigrators and scoffers, of whom there were a good many, that Guardsmen could not only physically fit into tanks, but that Guardsmen could also adapt themselves to the swifter-moving panorama of armoured warfare.

One of the first battalions to be transformed was the 2nd Battalion Irish Guards, then at Woking, and commanded by Lieutenant-Colonel G. L. Verney, MVO, who had recently taken over from Colonel Haydon. They had little equipment for training, and it had been a depressing winter, for they had been near enough to see the devastation caused by the blitz on London. The news that they were to convert to armour was at first received with incredulity, then with dismay. The first reaction was 'It will never work', followed by a burning determination to make it work. But there was little enough time to think about it for almost immediately the first bout of courses started and from then on it was courses and yet more courses – gunnery courses at Lulworth, tank-driving courses at Bovingdon, signal courses, vehicle maintenance courses, tactical courses.

At the beginning of September the first tanks arrived, four elderly Covenanters which had obviously been driven to death by their former owners. But they were the pride of the battalion and were regarded as show-pieces at first; however after a few days they were allowed to be taken, rather gingerly it must be added, for a little exercise round the park. One of the first moves was to name the tanks. Thus it came about that the Commanding Officer had 'St Patrick'; 'Munster', 'Leinster', 'Connaught' and 'Ulster' were given to the squadron commanders, who then chose place-names in Ireland for his remaining tanks, those with the initial letter 'A' for Number 1 Squadron and so on; from then on their tanks were, so to speak, blessed with life, and the crews became as attached to them as any sailor to his ship.

Perhaps the most notable occasion of those early days was a visit by the Major-General. It was decided to have an armoured review, and all four tanks were trundled out. The ceremonial experts were not entirely pleased; they felt that the parade would lack conformity and wanted to cut holes in the side of the tanks so 'the drivers could swing their arms'. But it was a fine march past; the tanks behaved perfectly, the guns were traversed and then dipped in salute, and only a few ill-natured spectators afterwards swore that rusty water had been seen to trickle from one of the gun-barrels.

The tanks continued to arrive, all second-hand to start with, but one day a brand-new one appeared, and with that the battalion reckoned that their existence had been properly acknowledged.

Inspections, exercises, and more training occupied them throughout 1943. In the meantime the 1st Battalion had been in action in North Africa and, later, in Italy. To replace their casualties the highly trained 3rd Battalion had been piecemeal chipped away and in six months had sent out no fewer than twenty-three officers and over 600 men. There were few who thought the 3rd Battalion would ever see action as a separate unit. But late in 1943 they were ordered to Malton, in Yorkshire, there to become one of the motorized battalions of the Guards Armoured Division.

They were desperately short of men; even with the injection of two full companies of Scots Guards, Right Flank and 'X' Company, their strength was still below par. To add to their problems Right Flank Company left before the winter was out. But by dint of monopolizing the reinforcements from the Holding Companies, then at Lingfield, by the spring of 1944 the battalion of the Irish Guards were also a viable and highly trained fighting force. They were in time to join the surreptitious movement of troops towards the South Coast in preparation for the invasion of Europe.

But the doings of the 1st Battalion Irish Guards in North Africa and Italy cannot be passed over, for these two campaigns were to see perhaps the finest feats and examples of staunchness, skill at arms and indomitable courage in the whole history of the Irish Guards.

Chapter 10

North Africa

The Battle of El Alamein began on October 23rd, 1942; by November 4th victory was assured and with this triumph the motif of the war in North Africa changed from one of attrition to one of pursuit, for Rommel was now in full retreat. Tobruk was recaptured on November 13th and Benghazi on the 30th. But before this, on the 8th, had taken place the long-awaited 'Operation Torch' – the Allied First Army's landing in French North Africa – apparently to the enemy's complete surprise.

A few days after the initial landings in French North Africa, 24th Independent Guards Brigade, which consisted of the 1st Irish Guards, the 1st Scots Guards and 5th Grenadiers, prepared for overseas service. They had been peacefully settled in southern England but now it was odds on North Africa being the ultimate destination. However, not until three months had elapsed were they actually under way.

In late February the battalion transport was ordered to move to Cardiff and Liverpool. A few days afterwards the rest of the battalion prepared to go by road to Glasgow.

At this stage there was a change in command. Lieutenant-Colonel C. A. Montagu-Douglas-Scott took over with Major D. M. L. Gordon-Watson, MC, as his second-in-command. To add a final disruption to carefully made plans, on their last exercise – a night attack on a hill, a prophetic note – the Regimental Sergeant-Major broke his ankle, and a truck from another unit ploughed into a rifle platoon and killed or maimed a number of men.

Eight liners were mustered in the Clyde for the Division. The battalion piled on board the converted P & O liner *Strathmore* where they lived like sardines. 'The people who arrange these things,' wrote Colonel Scott, 'never realize

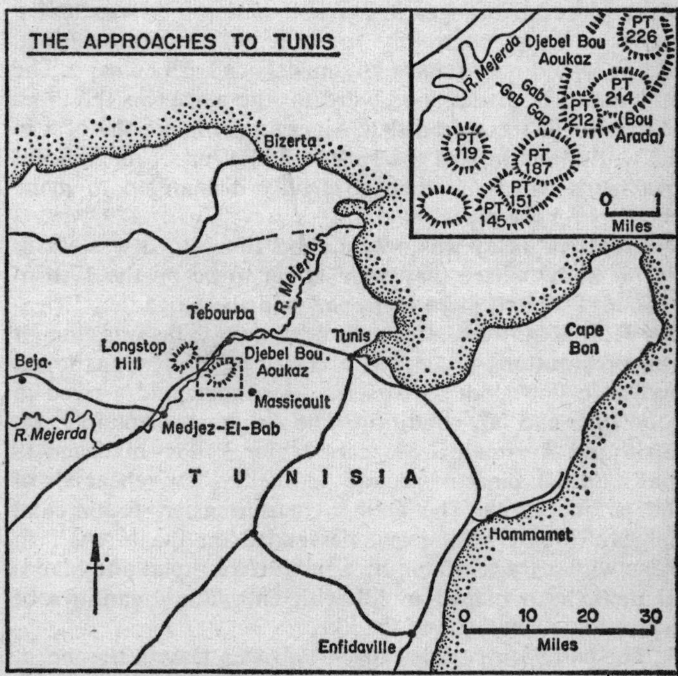

that two Guardsmen take up the room of three ordinary soldiers.' They sailed on March 1st.

The *Strathmore* docked at Algiers on the evening of March 9th, but it was getting on to midnight before the battalion reached their staging camp. Their arrival in the country was greeted by rain and some German aircraft – but the first was by far the more inconvenient as it turned their so-called camp into a glutinous wilderness.

'For the three days we were there,' wrote one strongly

disillusioned officer, 'it was rain, rain, rain, mud, mud, mud, mud always ankle deep, of the texture of paint, evil-smelling, fabulous, frightful, awful.' The French colonial settler in the nearest farmhouse did a grand trade in oranges. When the men had spent their few francs he charged a silver piece an orange; it did not matter if it was half a crown or a threepenny bit, you only got one orange for it, and he would not accept Holy medals as Irish coinage. The only man who had a good word for the place was the Pipe-Major, for it transpired that Algerian reeds are the best in the world, and he had the luck to find a hut of magnificent specimens which he then cheerfully dismantled to make chanters for the pipes.

St Patrick's Day was only a short time off. It was essential to know where they were going to be on the 17th of March. 'Probably here in Bône,' said the Brigadier, Brigadier R. B. R. Colvin, after their next move. This was enough and preparations proceeded at fever pitch. The shamrock had been flown out to General Alexander, and arrived in good time and fair condition. The Adjutant, Captain J. M. Park, and Regimental Sergeant-Major Pcilow managed to find a flattish piece of ground and held a few rehearsals of the parade itself. The acting Quartermaster, Regimental Sergeant-Major McKinney, descended on the NAAFI in Bône with a truck which, by a mixture of threat and blandishment, he managed to fill with enormous quantities of cigarettes, chocolate and the like.

'The battalion paraded on St Patrick's Day in the morning. It makes one proud to see them,' wrote the Brigadier to Regimental Headquarters. 'An American liaison officer with us was so moved when the battalion marched past that he blubbed freely.' With the shamrock had come a message from General Alexander. 'Welcome to the Micks. Now we'll get cracking.'

The subsequent dinner was deemed a great success. The warm and sticky red wine which had been substituted for beer was drunk like lemonade; too late did many discover that it had a hidden potency, and the town of Bône took full punishment that night. A week later Colonel Scott wrote, 'I am still getting reports from Bône. The terrible ruffians, even when they were being conveyed under escort, managed to pelt the Military Police with bully tins.'

But early the next morning they were off again – by road this time. Thankfully they slept off the excesses of St Patrick's Day – it was the last that most of the men in the 1st Battalion Irish Guards were ever to see.

'Proceed at once to Beja' – an agitated staff officer intercepted the battalion convoy in the middle of the afternoon – 'Tamara to the north has fallen, and Beja itself is now threatened.' The situation was said to be critical, but not apparently so critical as that at Medjez whither they were then rushed later the same day. In the course of twenty-four hours they came under command of three different divisions.

They were to stay in the lush Mejerda Valley for some time. It was beautiful country, of rolling corn, fat flocks and myriad birds. By day the ridges were held by only the Bren-gunners – the rest of the companies came back over the crest, waiting until dusk when all available men were required to prevent enemy patrols from slipping through somewhere along the barren slopes.

Gradually the battalion gained confidence in itself. The Commanding Officer had seen service in Norway. Major Gordon-Watson had also been present in the pre-war Palestine troubles, and most of the senior officers had been in action before. But the most experienced of all was the priest, Father J. R. (Dolly) Brookes, who as an Ensign had served under Alex in the First World War and remained with the Irish Guards until he took Holy Orders in 1925. The tall, imperturbable priest was universally loved and respected by the battalion, and it was to Father Dolly that one Guardsman said of the young officers who were all new to battle, as was indeed most of the battalion, 'Father, we thought they was all playboys until we saw them in action – after that we'd have followed them anywhere.'

It was not long before the battalion acquired a knowledge of what went on in the deceptively peaceful valley to their front, and on the unpleasant-looking ridge beyond, 'Recce Ridge' as it was known. They discovered that the forward slopes were swept by machine-gun fire from both flanks, that the lower slopes were infested with mines, and that the main German positions were just over the crest. They also found out that Arab informers usually managed to discover what was afoot behind the British lines and tell the Germans in good time.

This routine could not last. On March 18th the BBC announced that Eighth Army had attacked the Mareth Line 230 miles to the south, and they heard of the arrival of a new division somewhere on the right. But the Irish Guards were left alone with their small private war with the inhabitants of Recce Ridge.

Their preoccupation was interrupted on the 26th by the arrival of the Corps Commander, Lieutenant-General C. W. Allfrey. After being regaled with some red wine tasting strongly of petrol he gave an outline of how things were going elsewhere in Tunisia: Eighth Army were nearly through the Mareth Line; in the north First Army were about to strike a weakened sector; the Germans had no more reserves. Thus if the enemy were to help the other threatened sectors they would have to thin out in the centre, at Medjez. 'And this is where you come in,' he added, rather ominously everyone thought – it transpired that there was to be an attack by a single company on Recce Ridge. 'It was a nasty job,' wrote an officer afterwards, 'which entailed an advance across the valley in the dark, a climb up the slope, a battle on the ridge and then a withdrawal in daylight across the valley again.'

Punctually at midnight the first part of the approach route to Recce Ridge was cleared of mines by the assault pioneers and shortly afterwards Major S. J. R. Bucknill with Number 2 Company moved silently off, accompanied by two artillery officers to coordinate the massive fire support which had been allotted. The last man turned to wave and then Number 2 Company were swallowed by the night.

The watchers shivered and it was due only in part to the cold. The sky was clear and starry, there was no moon and no sound at all to break the stillness. It was wholly eerie.

Five AM and all was well so far. A faint whisper of 'Paddy Two, Paddy Two' had been heard at regular intervals on the wireless, otherwise there was nothing to report. It seemed ominously quiet to the small party of anxious watchers straining their eyes in the dark. The Commanding Officer had been brooding ever since the Corps Commander had paid his visit, and this change in him, so utterly out of character, had affected those around him. It was a sombre little group which peered hopefully towards Recce Ridge.

Then, at last, Major Bucknill came on the radio. All was fine for the moment; they were halfway up the slope, and apart from two German patrols which they had dodged without difficulty there was no sign of the enemy.

But no sooner was this message passed than they caught the sound everyone had been dreading. The crackle of machine-gun fire and the louder crump of hand-grenades could be heard across the valley. 'I want support on the eastern edge now,' the peremptory voice of Major Bucknill came urgently over the wireless. The artillery officers with him must have been calling for fire over their own radios at the same time for almost immediately the gunners answered the call, and with a noise like thunder the barrage broke about Recce Ridge.

It was soon light enough to see the far slope, but all the watchers could spot was the occasional figure moving around on the near side of the crest. The noise of small-arms fire still carried across the valley, but the wireless was completely dead and anything might be happening.

Colonel Scott called for smoke, and soon canister after canister was raining down on the ridge. The pall of smoke brought a wild burst of firing, but it died away as the wind dispersed the cloud into wisps. The firing was more intermittent now – no Brens, only the occasional rifle shot answered by the faint but distinctive rattle of the German Schmeisser machine-pistols. And then silence. 'It's all over now, one way or the other,' said Colonel Scott. 'Keep up the smoke as long as possible.'

Five wounded men came back from the left, and later the following night two, unwounded, from the right. But that was all out of 103 officers and men of Number 2 Company 1st Battalion Irish Guards. For it transpired that the Germans had been expecting them for days – although never in their wildest dreams at only company strength.

Number 2 Company was replaced in its entirety from the first-line reinforcements held in camp near Algiers. The new arrivals were immensely relieved to find anything left of the 1st Battalion, for there had been dreadful rumours in the rear areas.

Recce Ridge was captured some days later by a battalion of the 78th Division, with Churchill tanks in support.

With the subsequent withdrawal of the Germans in front of Medjez the stage was set for the final offensive. The enemy were now occupying two fronts at right angles to each other. On the southern flank, on a line roughly running west from Enfidaville, they opposed Eighth Army; along their line running northwards to the shores of the Mediterranean they faced First Army. General Alexander decided to attack their western flank, and by breaking out towards Tunis split the enemy in half. The crucial area was around Medjez-el-Bab. Once tanks were in the plain beyond Massicault, Tunis would fall and the whole German position in North Africa would inevitably collapse.

But before the armour could break out the Germans had to be cleared from around Longstop Hill and the Bou Aoukaz positions (see map, p 121). This was infantry work, and 1st Division allocated the unenviable task to 24th Guards Brigade.

The Germans mounted a strong spoiling attack with their Hermann Goering Division twenty-four hours before the assault was to go in. But thanks to the timely intervention of some Churchill tanks of the North Irish Horse –

personally brought forward by Brigadier Colvin – they were beaten off with heavy loss.

The 24th Guards Brigade attacked on the night of April 23rd – Good Friday. After a day of confused fighting the 1st Scots Guards reported themselves firm on Point 145, the first of the three hills which protrude into the plain short of the Bou Aoukaz itself and the possession of which is a necessary prelude to any attack on the main feature. The other two hills, Points 151 and 187, were taken by the Irish Guards on the night of Easter Day. The springboard for the attack on 'The Bou' itself was now ready.

Easter Monday, April 26th, was oppressively hot and very uncomfortable. The companies burrowed as best they could into the hard, unyielding earth, striving to gain a little protection, for they were in full view of the enemy across the Gab-Gab Gap and from 'The Bou' beyond. An attack had been intended for that night, but time was too short for it to be mounted. The 27th was chosen – at last light, which was at 18.30 hours. (See map, p 128.)

From the forward edge of Point 187 the nearest feature of 'The Bou' was about a mile away. The invervening Gab-Gab Gap was covered with waist-high corn interspersed with an olive grove to the immediate right of Point 187, and another further on and to the right of the first hill of 'The Bou' proper. 'The Bou' itself was a long, narrow ridge with a pimple on each end, Points 212 (Bou Arada) and 214; beyond was the principal massif of Bou Aoukaz.

By dint of some judicious squirming the company commanders could see their objectives, or some of them. It was an uninviting task, for they were advancing into a funnel with dominating heights on both sides. The undulating corn looked deceptively good cover, the olive groves might afford a little protection, but otherwise the Gab-Gab Gap was flat, open and completely overlooked to the north-west, the north and the north-east; and the further north they went the more overlooked they would become. With this unpleasant prospect before him the Commanding Officer chose the only reasonably protected approach, and picked the olive groves as his axis.

Colonel Scott still considers that the attack as planned was a feasible operation of war, although highly risky. A

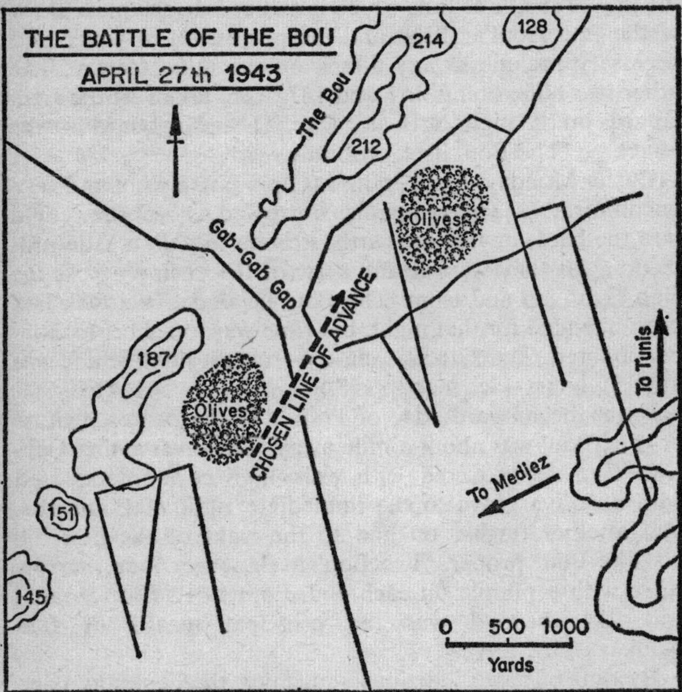

night attack would have a fair chance of success, but a daylight attack which they were then ordered to do was pure suicide. For at midday they were appalled to hear that zero-hour had been brought forward to 1600 hours. This meant that the attack would be in broad daylight, in the heat of the afternoon and across an open plain dominated on all sides by the Germans. There were reasons enough for the change of plan. But there wasn't a Mick in the 1st Battalion Irish Guards who did not know and fully comprehend the fate and carnage which lay ahead.

The battalion formed up under the lee of a little hill shortly before zero-hour; four rifle companies, each of about ninety men. At four o'clock, Number 3 Company leading, they moved off into the open, 'stepping a long, steady pace'. There were only a couple of hundred yards to go to the nearest cornfield, but before the first man reached it the German guns had opened up. 'They threw everything but their cap-badges at us,' related one man; guns, mortars, and those abominable machines the six-barrelled mortars, everything within miles, let drive. The platoons spread out into open order and plunged into the waist-high corn. The fire intensified and heavy machine-guns joined in; soon the whole cornfield was ripped and torn. Part of it was burning smokily. Amid the tall poppies that stood out over the corn there sprang up a new crop – rifle-butts. They appeared suddenly and so thickly, that it was almost a surprise to look beyond them and see the thin line of men plodding steadily on towards the olive grove. 'We could not believe it,' said a German prisoner afterwards; 'we thought nobody could cross that plain.' 'Thank God for drill,' said one Guardsman, 'it keeps you going.' They fixed their eyes on the olive grove ahead – it looked so peaceful and safe.

Within minutes of the start two company commanders were out of action, Captain J. B. FitzGerald killed and Captain J. H. Powell-Edwards wounded. The thin line was getting thinner all the time.

But there was no hesitation. The companies reached the first olive grove, which was a registered target, and attracted the full weight of the German fire-power. By now three of the four rifle companies had only one officer left, and shortly afterwards the Regimental Sergeant-Major was killed.

As soon as Number 4 Company emerged from the sparse cover of the grove they lost their only surviving officer. Captain O. Chesterton, commanding Number 1 Company, although wounded twice himself, managed to absorb the remnants of the other company into his own and then contacted the Commanding Officer over his wireless. 'We will be able to take Point 214 as soon as it is dark. Can we wait till then – about half an hour?'

The original plan had been for two companies to make for Point 214, the furthermost of the two pimples, while the other two headed for Point 128 to the east and isolated from the main massif. But it was discovered that Point 128 was insignificant and anyway was unapproachable in daylight, so Colonel Scott decided to switch the whole battalion, or what was left of it, to Points 212 and 214 instead.

As darkness fell, the advance continued. Number 1 Company plodded on up the hill, bayonets at the ready, but the Germans didn't wait and fifty or sixty were seen scurrying away to safety. By every means possible the remnants of the 1st Battalion Irish Guards were scraped together – by midnight 173 men of the four rifle companies and advanced Battalion Headquarters were on Point 212.

The rest lay down in the corn or the olive groves calling to the stretcher-bearers, or silent, and marked only by a rifle-butt silhouetted in the glow of burning Arab huts.

The top of the ridge was bare rock, with only a few pockets of earth. The men worked all night digging, hacking and shifting boulders to get the slit-trenches built before dawn. The carriers ran loads of ammunition, rations and water from the battalion's old position to as near as they could to the foot of the ridge and gave lifts back to the walking wounded. Father Dolly Brookes, with a team of 'beaters', walked backwards and forwards over the plain and along the course taken by the companies. He still had with him three young Germans who had surrendered to him some time in the afternoon as he was walking along swinging his blackthorn stick; they were an embarrassment till he turned them into stretcher-bearers. The ambulance came up through the Gab-Gab Gap and collected the wounded off the track. By dawn they were cleared from the plain – the dead would have to wait.

Elsewhere the brigade attack had had mixed success.

The 5th Grenadiers had succeeded in securing their objectives to the west without difficulty; but the Scots Guards had been held up short of the massif of 'The Bou' itself.

So the survivors of the 1st Battalion Irish Guards found themselves isolated, forming a salient of their own and unable even to get their support-weapons forward as the slope of Point 212 was too steep for the carriers, and the anti-tank guns were too unwieldy. Of greater importance was the fact that the Number 19 wireless set, the only link with Brigade Headquarters, could not be brought up either – so this decided Colonel Scott to pull back his main headquarters to Point 187 on the other side of the Gap. It was as well he did, for half an hour after they moved the Germans took charge of the Gab-Gab Gap and the remains of the battalion were now completely isolated.

They took stock of their position: they had five officers left between what had once formed four rifle companies and advanced Battalion Headquarters, but no warrant-officers and only one senior NCO. Otherwise the numbers in each company were more or less even, so they split the ridge into company groups. They had managed to bring up twenty boxes of ammunition; there were three or four sacks of bread and a dozen jerrycans of water or tea – that was all until night fell, and perhaps longer. 'On that morning,' the War Diary says, 'April 28th, 1943, 173 Irish Guardsmen, armed with Brens, rifles and two grenades each, held the hill. On Friday night, April 30th, after five large attacks, 80 Irish Guardsmen, armed with Brens and rifles, still held it.' At the end of it all the Adjutant, Captain D. J. L. FitzGerald, wrote a report from which the following are extracts:

'Wednesday, April 28th. After a quiet morning the enemy began to mortar the ridge heavily about midday. The only result of this was that any slit-trenches that were not finished were rapidly deepened. At about 1400 hours, 88-mm guns and tanks, firing high explosives, opened up on the crest from close range in the olive groves behind us. This fire was heavy, most accurate and most unpleasant.

The Germans had clearly set to work to "soften" the defences. The men were withdrawn to the top of the reverse slope, leaving on the eastern side only look-outs dug in as well as possible. At 1500 hours the fire intensified. Salvoes from six-barrelled mortars screamed through the air like Green Line buses and crashed in quick succession along the ridge. The look-outs shouted "Here they come!" German infantry were climbing the west slope up the end of the ridge nearest Point 214. This was the first opportunity the Guardsmen had of engaging the enemy personally, and they took it eagerly. Numbers 3 and 4 Companies – we still called the small groups companies – moved up to the crest, and as soon as the shelling slackened, halted the Germans halfway up the hill with rifle and Bren fire. When the Germans were already wavering a section of Number 1 Company came over from the right flank, which was not engaged. The increase in the volume of fire had a splendid effect. The Germans turned and ran back down the hill. We were astonished at the sight, but soon recovered. The Guardsmen went forward and shot down the fugitives till they disappeared into the corn below. Two German officers who tried to rally their troops were picked off by rifles. The whole force was elated by this success, and from then on morale was at the highest possible level. The Guardsmen never for one moment doubted their ability to thrash and to go on thrashing the Germans as long as the ammunition and water held out.'

After that attack the shooting and mortaring began again in earnest and continued more or less non-stop until the Friday night. But worst of all was the thirst. With no protection from the blinding sun and the air hazy with the dust from exploding shells, the little water they had proved nothing like enough. At night the men ventured down the slope to take the water-bottles from the German dead, but by day they could only suffer.

At six o'clock that evening (Wednesday) the Germans began to work round the left of the ridge and reached a gully below Point 214. Clearly another attack was imminent; duly at seven they started again, and despite fierce resistance, managed to get to the top of Point 214. Since life would have been intolerable if the enemy had held this point, Lieutenant C. D. Kennard – who was awarded an immediate DSO after the battle – led a bayonet and Tommy-gun charge and the enemy was swept off it and grenaded down the hill. The Germans continued to press their attacks on the northern half of the ridge and nearly reached the flat top. The report continued: 'It was getting dark and the situation looked serious. At our moment of greatest need Sergeant Musgrove and his 3-inch mortar team came running along the reverse slope, carrying their mortar. The mortar had had to be abandoned that morning at the southern foot of Point 212. Sergeant Musgrove had quietly rescued it in the half-light. It was assembled in record time and all twenty bombs were shot over the ridge at minimum range, with alternative switches to left and right. The effect on the enemy could be judged by the screams which followed each burst. Number 4 and Number 1 Company then ran forward and destroyed the leading Germans, who had been cut off from the main body by the mortar fire. If we had had more grenades we could have done even more damage. As it was, the Guardsmen finished by throwing rocks after the Germans . . .'

The Germans retired, and throughout the night they could be heard digging halfway up the hill and collecting their dead and wounded. The roll of one German company was called by what was described as 'a very bad type of CSM' and everyone was delighted to note that he was deficient of a good half of the names. The wounded who could walk were sent down the western slope to get back as best they could. It was at least a chance of getting their wounds treated before they went gangrenous, for, as Guardsman Hickey, the senior stretcher-bearer, said, 'I have no time for the gangrene.'

Dawn on Thursday the 29th passed relatively peacefully; there was no ground attack, but the shelling and mortaring never stopped. Not until nine o'clock that morning did the enemy come on again. As the report said, 'This time it was in even greater strength, and all along the ridge from 212 to 214, including both Points. The first warning came from the look-out man for Point 212. A German company was forming up for the attack just below and by leaning over the top it was possible to see them assembling. Lance-Corporal Kenneally decided that this was the right moment to attack them himself. He charged down the bare forward slope straight into the Germans, firing his Bren-gun from the hip as he went. The Germans were thrown off balance by this extraordinary assault. Having cut right through them, Kenneally turned and began shooting the men behind him. It was too much for the Germans. They started firing in all directions at once, trying to follow their assailant as he leapt about the hillside. They were still firing at each other when Number 1 Company came forward and finished them off. Number 4 Company dealt with the attack in the centre in a more conventional way with the help of most of Number 1 Company which, thanks to Lance-Corporal Kenneally, was now free to move over ... After a few hours' respite another attack was made by fresh German troops. By this time we were getting quite used to them and as usual drove them off. We watched and waited eagerly for the Germans to waver under the fire; at the first sign of it the Guardsmen jumped out of their holes and rushed at them. It was a highly successful policy, and made a nice change, as Guardsman Foran said.' CQMS Mercer led Number 4 Company into a very neat charge, with himself on one flank and Sergeant Lynch on the other, 'to keep the dressing!'

They were joined here by Major H. L. S. Young and Lieutenant B. T. Synge with a Vickers machine-gun and team, nicely in time to have a grand shoot at yet another attack.

On the Friday morning two more officers arrived, having thumbed a lift in a carrier bringing up rations. And then, at eleven o'clock, the Germans made their biggest and last attack of all. It opened with a violent bombardment from

88-mm, tanks, field-guns and the damnable six-barrelled mortars. The 88-mm and tank guns tried to knock the top off the ridge with HE and solid shot, while the field-guns and six-barrelled mortars pounded what used to be the reverse slope. Then came the German infantry in exactly the same way as before, but in greater numbers and with greater persistence. There was a higher percentage of officers than they had ever seen with any troops – half were in front leading ('these were easily picked off') and half were behind, driving on their troops ('these were harder to get at'). The assault on Point 212 was broken up by Lance-Corporal Kenneally and a Sergeant Salt of the Reconnaissance Regiment, who with his troop had found his way forward to 'The Bou' before the main German attack.

The words of the citation awarding him the Victoria Cross read: 'Lance-Corporal John Patrick Kenneally repeated his remarkable exploit on the morning of April 30th, 1943, when, accompanied by a sergeant of the Reconnaissance Corps, he again charged the enemy forming up for an assault. This time he so harassed the enemy, inflicting many casualties, that this projected attack was frustrated. The enemy's strength was again about one company. It was only when he was noticed hopping from one position to another further to the left in order to support another company, carrying his gun in one hand and leaning on a Guardsman with the other, that it was discovered that he had been wounded. He refused to give up his Bren-gun, claiming that he was the only one who understood that gun, and continued to fight all day with great courage, devotion to duty and disregard for his own safety . . . His extraordinary gallantry in attacking single-handed a massed body of the enemy and breaking up an attack on two occasions was an achievement that can seldom have been equalled.' The gallant Sergeant Salt was killed shortly afterwards.

On the left the Germans actually got over to their side of Point 214 and flooded along the top of the ridge. Shepherded by tanks, they fell on Number 2 Company. The Guardsmen rose to meet them and they fought it out on the edge of their trenches. 'What with tanks sliding around on the rocks and the Germans shouting their heads off, it was a real "fracass",' said one man. Two sections were over-run and led away, but one of them took an opportunity to outwit their captors and rejoined the battalion after twenty minutes' absence, for which the section commander asked to be excused. (The others later escaped from a prison ship when it was bombed, and reported for duty in Tunis.)

In the confusion the rest of the force had a moment to organize a counter-attack. A platoon sergeant led Number 3 Company against Point 214, where Lieutenant Kennard came careering through them. Lieutenant D. G. Madden gathered Number 4 Company and hurled himself and them along the ridge. These charges shattered the Germans, and the whole line swept forward cheering and shouting 'Up the Micks!' 'We chased the enemy halfway down the slope, firing "into the brown", and bayoneting the less fleet-footed. I have never seen anything like it, and it was with great difficulty that the men were restrained from going down into the cornfield,' wrote the Adjutant.

And that was all. The enemy had had enough and over 700 German dead were later counted in the area of 'The Bou'. A company of Grenadiers fought their way to join the Micks in the early afternoon, and at about 1800 hours another of the Gordons joined them on 'The Bou'. The Adjutant's account ends: 'At 0300 hours, Saturday, May 1st, we were relieved. 173 men reached Points 212 and 214 on Tuesday night; 80 left it on Saturday morning, which included those men such as the machine-gun section, who managed to get up during the siege. The conduct of the battalion throughout these days was beyond all praise. They were short of food, short of sleep, short of ammunition, very short of water, continually shelled, mortared and machine-gunned, sniped and often surrounded; but never did I see anything but a fixed determination to hold our positions – for ever, if need be – a complete contempt for the German infantry, and a solid and cheerful endurance of the German fire. After the repulse of the first attack morale was at the highest, and I have never seen the Guardsmen more cheerful. There was no suggestion of complaint of any kind, and the most dangerous and uncomfortable tasks were carried out almost before the order was given. Very few orders were given by me or anybody else – there was hardly any necessity, for the whole force acted as one cohesive and intelligent body. Every man, including those with the longest "sheets", did his duty, and more than his duty. I always knew the battalion was good, but I never guessed it was so good, and personally I shall never forget their courage and uncomplaining endurance, their unfailing cheerfulness, the silence and patience of the wounded and the kindness and gentleness of the men to their wounded comrades and the spirit of cooperation among themselves. As they said themselves, "We are all Micks together, and the Micks muck in." It was an honour to be with such men.'

A great white cross now stands on Hill 212. On it is inscribed: 'To the memory of the Officers, Warrant Officers, Non-Commissioned Officers and Guardsmen of the 1st Bn Irish Guards who died on and around this hill April 27th–30th, 1943. QUIS SEPARABIT?'

After the battle of 'The Bou' the battalion, now only one company strong, were withdrawn and they watched the subsequent proceedings with relish. On May 6th the massed might of First Army, preceded by a tremendous air effort, smashed their way to Massicault, halfway to Tunis. On May 7th, 6th and 7th Armoured Divisions took up the running. After a few brisk skirmishes opposition collapsed – they were through. At 11.45 AM Tunis fell. And on May 10th, 1943, General Alexander signalled to the Prime Minister: 'Sir, it is my duty to report that the Tunisian campaign is over. All enemy resistance has ended. We are masters of the North African shore.'

The Irish Guards shuffled forward behind the advancing troops but took no part. From time to time they welcomed back those lightly wounded who had been able to escape from the evacuation authorities, but they were still mentally reeling from the hammering they had received. Consolation came from two quarters. First, from the Divisional Commander: 'Now that this particular phase of the operations is over,' he wrote, 'may I express my very great admiration for the gallant conduct of the 24th Guards Brigade. The 1st Division was selected to bear the brunt of forcing an entry through the crust of the enemy to enable the armour to break through. All three brigades had very strong enemy positions to attack which they did most gallantly, but the relentless courage, the cheerful sacrifices and the great tenacity of the 24th Guards Brigade was outstanding, and indeed without it the victory of First Army could never have been achieved. While it is impossible to differentiate between the conduct of all three battalions, I think the story of the Irish Guards on Hill 212 will always stand in red letters on the pages of that glorious Regiment's history. Your losses were great and terrible, but my heart goes out to you in thankfulness that such courage should produce a reward, the true value of which, at this time, no man can assess.' The second tribute was from General Alexander himself: 'Heartiest congratulations to you and all ranks of the battalion for your magnificent fight, which has not only added fresh laurels to the illustrious name of the Regiment, but has been of the utmost importance to our whole battle. I am immensely proud of you all; I am very sorry about your losses.'

The Victory Parade at Tunis involved as much preparation as a major battle. The battalion crowded down to the Avenue Jules Ferry and climbed on the tanks of the North Irish Horse to get a view of its own contingent marching past. The highly polished, neatly dressed Guardsmen were followed in the parade by a horde of Moroccan Goums, wearing dressing-gowns and slippers, and then by a regiment of the French Foreign Legion, who had fought alongside the battalion in Norway. On the first free Sunday Father Brookes sang a High Mass of Thanksgiving in Carthage Cathedral for the battalion and the 78th Division, which included an Irish brigade. When the sound of the pipes which were leading the marching contingents drifted into the great, still basilica, one of the white-robed priests sprang from his stall and rushed down the aisle crying, 'I must see the boys, I must.' His name in religion was Père Angelicus, but he had been born Flynn in County Westmeath. He later met members of the 1st Battalion in Jerusalem during the time they were stationed in the Canal Zone in 1955, and many still remember his kindness and hospitality.

Three months of sweating peace followed at a camp four miles from Tunis. Rudimentary field training succeeded drill parades, and gradually the reinforcements who had come from England were absorbed into four newly emergent rifle companies.

At the end of August they were ordered to Hammamet on Cape Bon for field training. It was at this time they learned that Lance-Corporal (now Sergeant) Kenneally had been awarded the Victoria Cross.

Letters and telegrams of congratulation poured in; everyone who had ever had anything to do with the battalion wrote or called, and a stream of mail followed them down to Hammamet. On August 27th General Alexander made the presentation. 'A tremendous day,' wrote Colonel Scott. 'General Alexander arrived, accompanied by the Corps Commander, the Divisional Commander – Major-General G. W. R. Templer, DSO, OBE – Lieutenant-General Bradley of the US Army and Major-General Lemnitzer, the American Deputy Chief of Staff. The battalion were formed up on three sides of a square in a rather dusty field and under a very hot sun. The place was crowded with reporters and photographers. We had no idea what sort of a parade General Alexander wanted, but he said on arrival that he wanted to go round the whole battalion. He did this, and then Kenneally was marched up to him through the battalion by Regimental Sergeant-Major McLoughlin. After he had pinned the Cross on Sergeant Kenneally he made a speech to the battalion, congratulating them personally on having played a vital part in his victory, and probably in total victory. Then the battalion marched past while he stood at the saluting base with Sergeant Kenneally on his right.' Sergeant Kenneally now lives in Warwickshire.

Meanwhile the war swept on. Sicily was invaded in July; on September 9th the landings at Salerno took place. But still no move for the 1st Division. Every rumour proved false, every warning order – and there were several – was cancelled.

On December 4th they piled aboard the *Llangibby Castle*. The battalion were adept at circumventing tiresome officialdom by this time and the embarkation went smoothly enough after an initial complication caused by an over-zealous American staff-officer who insisted on checking each man against his nominal roll, which was in alphabetical order whereas the battalion was not. 'I don't doubt your word,' he told the Adjutant, 'but it is the routine.' The Adjutant spoke to the Company Sergeant-Majors, the Company Sergeant-Majors spoke to the companies, and the 'routine' was followed. As the names were called out the battalion filed up the gangway by platoons, answering to any name. Embarkations did not normally go as smoothly as this, in the American's experience, but whenever he called back a Guardsman to confirm his name he was always satisfied. 'What's your name?' 'What name did you call, sir?' 'The five O'Donnell boys.' 'I'm one of them, sir.'

The convoy sailed the following morning. The battalion crowded on deck to see the last of Africa. Someone began to sing his own version of *Galway Bay* and the battalion took it up, repeating fervently the verse that begins, 'Oh the wind that blows across the sea from Tunis', and ends:

And the Arabs gathering cactus in the hedges
Speak a language we have no desire to know.

Chapter 11

Anzio

The Irish Guards Official History notes that 'the words "what might have been" are the saddest in life; they are the keynote of the Anzio landing'. Nothing has come to light in the years since this was written to refute the judgement. Some feel that Anzio was the greatest missed opportunity in the history of the Second World War; it certainly provided the setting for some of the most bitter infantry fighting of the war.

The first half of January had been spent by the 1st Battalion Irish Guards in the town of Gragnano, a town near Naples, noteworthy chiefly for the huge empty factories which served as excellent makeshift barracks.

During this time they had managed, by dint of much persuasion, to get hold of their Regimental Band who were touring Italy, and the Irish Guards marched out of Gragnano to the Regimental Quick March *St Patrick's Day*. They swung down the hill into Castellammare. The brave sound of *The Minstrel Boy* and *The Wearing of the Green* and the steady tread of Guardsmen rang through the shabby streets. Italians lined the pavements and crowded the main square, clapping and cheering 'Viva, viva!' as if the whole thing was for their benefit – as indeed, to a certain extent, it was. Colonel Scott, standing on the pedestal of a statue, took the salute. The battalion embarked on their allotted transport and the convoy sailed out of Naples Bay.

As dawn lightened the sky on the morning of January 22nd the Guardsmen tumbled on deck to be greeted with an astonishing sight; a peaceful landscape, a lovely dawn, and no Germans, not even the usually ubiquitous reconnaissance planes; it was all very unreal.

The 1st Scots Guards were ashore in the first wave, the remaining battalions of 24th Guards Brigade, the 5th Grenadiers and 1st Irish Guards, followed at seven o'clock. The battalion landed dry-shod and walked inland. They passed Italian farm labourers on their way to work and greeted them politely. The Italians showed singularly little interest. '*Dove Tedeschi?*' asked the Guardsmen. '*Niente Tedeschi,*' replied the Italians, and jerked a thumb vaguely northwards, '*Roma.*' 'Rome' translated the scholars. 'We will give the Holy Father a holiday,' said one enthusiastic man, 'and make Father Brookes acting-unpaid Pope.'

That night it began to rain and it was bitterly cold; the morning was not much better. But permeating all their thoughts was a terrible creeping disquiet that a wonderful opportunity was being missed.

At last, as dusk was drawing in on their second night on the beachhead – thirty-six hours after they had landed – the Grenadiers received orders to make an 'armoured reconnaissance to the village of Carroceto and beyond'. They were to start at dawn.

Guarding the southern approach to Rome lie the Alban Hills, a towering mass of mountain, and a natural fortress of formidable strength. South of the hills lies the Campo di Carne – the Plain of Flesh, a name of bitter interpretation before long – a flat expanse stretching southwards to the sea at Anzio, twenty miles away. Once an impenetrable and malarial bog, by dint of hard work over the years it had been transformed by Mussolini into an outwardly magnificent farming area. But the water was never far from the surface, and in wet weather the place rapidly became a bog again, as the men at Anzio soon found to their cost. In the centre of this area was the little town of Aprilia. Previously an insignificant collection of houses, it had been transformed by the Fascists into a town-planner's paradise – shops, offices, a church, a playground, the whole dominated by the Fascist Party Headquarters with its tall tower. It looked from a distance like a factory chimney, and it was known as 'the Factory' by those who fought at Anzio.

There is a main road and railway running due north from Anzio. Six miles from the little port a concrete bridge

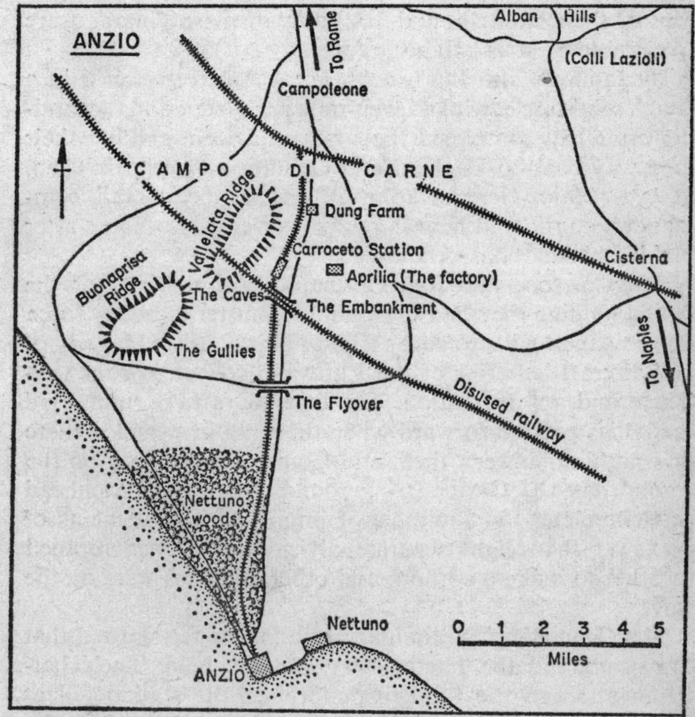

carried a transverse road over the two – this was known as the Flyover. Between the Flyover and the next feature, called the Embankment, lie a bewildering succession of gullies – deep, muddy, heavily overgrown channels often twenty to thirty feet deep, with sheer sides and bramble-covered banks which stretched right up to the road and railway. Beyond the Embankment lay the Factory, and two miles further north across flat, featureless country was the farm which became known as Dung Farm. It was four miles from the Flyover to Dung Farm. In the course of seven weeks, along this length and to a width of little more than a mile on either side of the road and railway, the three battalions of 24th Guards Brigade were to lose eighty per cent of their strength, and 1st Battalion Irish Guards were to cease to exist as a fighting force.

By January 24th the beachhead at Anzio stretched in a rough semi-circle, with a seven-mile perimeter; and the landing craft had completed their task and were making their long way back to England and Operation Overlord. Then, at last, Major-General John P. Lucas, the overall commander, considered he was ready – seventy-two hours after making his unopposed landing.

And so, too, were the Germans. Within hours of the Allied landing they had been able to muster a motley force on the Alban Hills to cover Rome. This consisted largely of anti-aircraft batteries of the Luftwaffe collected from anywhere and any formation. The first hours were vital. Had the Allies pressed forward when they had first landed there was nothing between them and Rome. But by noon on the second day the German ring round the Allied beachhead was complete. By a miracle of improvisation, elements of no fewer than eight separate divisions had been scraped together to make a cordon; and other divisions were on the way.

The Grenadiers' preliminary reconnaissance showed that Carroceto and the Factory were lightly held. Their battalion attacked the following morning; the station fell at eleven o'clock and the Factory at half past two. The 1st Irish Guards were ordered up to take over the left flank and the station.

There they found themselves at the point of a very thin salient. To their right lay the Grenadiers in Aprilia; to their rear were the Scots Guards; to their left a regiment of the German 3rd Panzer Division; the approach was down the main road, in full view of the Buonoprisa and Vallelata ridges, both dominating their own positions and both occupied by the enemy; and in the distance, to the north-east, was the Colli Lazioli, the first of the Alban Hills. To be on that side of the beachhead at all was later described by one American soldier as being like 'a bacterium in a bottle', and few of those who survived would disagree, for this was the first taste of that terrible feeling of being overlooked, and being able to do nothing about it.

The battalion were well established by nightfall of January 25th, and apart from the constant shelling and occasional sniping, things had been comparatively quiet. As darkness came so did the rain, and they spent a bitterly cold, soggy night, enlivened only by the hail which woke up everyone a few hours before dawn.

As the forward company, Number 3, were eating their sodden breakfasts came the first call for action. Captain D. M. Kennedy, the Company Commander, reported nine enemy tanks in the mist less than a mile away.

No sooner had Kennedy reported the tanks than the Germans opened up a fierce concentration on the battalion. Guns of every kind, calibre and nationality – German, Italian, Czech, French, Russian, Yugoslav guns, modern 105 and 220-mm, strange old field pieces and monster railway guns – opened up, and from the higher ground round the saucer which was the Campo di Carne the Germans could watch every shell land.

The tanks rumbled closer. Number 3 Company's anti-tank guns fired for the first and last time – two tanks went up in flames and then the guns disintegrated as the Germans took retaliatory action. Now it was the turn of the German infantry. Through his glasses Captain Kennedy saw a herd of them advancing down the Anzio railway, two companies up and one in reserve. It was a target fit for the guns. The Germans came on through the first few ranging rounds, but then completely disappeared into a welter of explosions as the main barrage of the guns took effect. Within moments there was neither sight nor sound of the German attack.

Rear Battalion Headquarters, sheltering in a farmhouse, was now receiving its share of attention. The first salvo from a batch of over-curious enemy tanks had killed one Guardsman and wounded the Sergeant-Major and two others. Regimental Sergeant-Major McLoughlin reported to the Adjutant. He had cut away the sleeve of his battle-dress jacket and was firmly grasping his massive forearm. The blood trickled through his fingers. 'It's nothing, sir, really it isn't,' he said. 'Why did I have to be hit so soon? Don't let them evacuate me.' His square face, with its fierce black moustache, was as disappointed as the face of a child cheated of a pantomime. Three days before the battalion left Scotland for North Africa, Regimental Sergeant-Major McLoughlin had broken his ankle on a night exercise. He rejoined after the Tunisian campaign. His ambition in life was to fight with his battalion, and now he was wounded. The Adjutant watched him trudge wearily to the Regimental Aid Post. 'If such a thing were possible, I'd say the Sergeant-Major is crying.'

The mist had completely cleared by now, to give the Germans an uninterrupted view of the battlefield. It was now a question of waiting and Father Brookes took the opportunity to read his office. He paced delicately up and down the track behind the embankment, a steel helmet perched on top of his cap, and his nose buried in his breviary. It was a very comforting sight.

The seemingly interminable day at last came to an end, and with that the moles of the 1st Battalion Irish Guards came out of the ground to take advantage of the mist which now settled over the Campo di Carne. Fires of all sorts, vehicles, tanks or farms, studded the countryside – a few more would make no difference, so the companies set about preparing themselves a hot meal. The Adjutant sat by some glowing embers checking casualties. Largely from shelling alone the battalion had lost over 120 men of whom thirty were dead. So ended what Father Brookes, a veteran of the First World War, described as the worst day he had ever spent under fire.

At dawn on the 26th the shelling began again, and there was no covering mist this time. Rumours abounded of a big armoured thrust, and a bevy of American aides appeared demanding to know where they could deploy their tanks. The Americans didn't believe it when told there was nowhere – until the afternoon, by which time they had proved it for themselves at some cost.

The tanks had failed, and now it was the turn of the infantry again. The Scots Guards were ordered to take and hold Dung Farm, prominent a couple of miles beyond the station. This they succeeded in doing, but a subsequent follow-up by the Grenadiers ended in disaster, for their Order Group missed the inconspicuous turn to Dung Farm and drove straight into the German lines. Three officers were killed and four more wounded and captured, including all the company commanders. To let the Grenadiers regroup after this shattering blow, the Irish Guards were told to take on the night attack instead.

The plan was for the Scots and Irish Guards to advance side by side: the Scots Guards up the general line of the road, the Irish up the parallel line of the railway to as far as the railway bridge a mile or so short of Campoleone Station. The left flank would be open for the duration of the advance, so the task of clearing it was given to Number 3 Company and some tank-busters. The shelling had now ceased. It was a silent night, and very dark; moonrise was not for several hours. Quietly the Irish Guards companies moved off; Numbers 1 and 2 up the main road, while Number 3 veered to the left and their own specific task of flank protection. At the start-line they spread out, but there was no wait, for at that moment their own barrage opened up – unfortunately some of it short and on the start-line itself – an unnerving experience at best. Following it closely they floundered their way forward over, into and out of the drainage ditches which impeded their way.

'Halfway to the objective and no opposition' was the report from the two leading companies; Number 3's wireless seemed to have broken down and there had been no contact since they had departed, but all was quiet to the left, so presumably all was well.

Then the Germans opened up. A machine-gun started firing upon the left of Number 1 Company, and another on the right of Number 2. More joined in, flares were sent up and shells and mortar bombs came down. There were now six machine-guns in a line on the narrow stretch of ground between the railway and the road. They shot a solid curtain of bullets into Number 1 Company, who suffered crippling casualties. Number 2 Company was also in severe trouble, but to their left it seemed as if Number 1 Company had by nonw found an easier way, so Major G. FitzGerald, commanding Number 2 Company, decided to switch his axis.

Number 1 Company, under Major Sir Ian Stewart-Richardson, had managed to find some drainage ditches conveniently running north–south, and reached their objective of the railway bridge with only one further casualty and a few alarms. Here they were joined by Major FitzGerald with his company, and as they only had one wireless set working between them – and that soon broke down – the two company headquarters joined forces.

The full peril of the position of the two leading companies was now becoming apparent. To their left there was no sign of Number 3 Company, but a lot of Germans instead; to their right were more Germans – although the Scots Guards had managed to fight on to the extreme right-hand edge of their own objective. Ahead lay a strong defensive position, or so it seemed from the amount of fire; and from the hair-raising experience of the support platoon commanders trying to reconnoitre a route for the anti-tank guns and tank-busters forward from Battalion Headquarters, it was also clear that a good number of Germans had been by-passed in the dark and were only now beginning to wake up. If dawn came and the two isolated companies possessed no anti-tank gun protection it would be merely a question of time before they were mopped up by German armour. For the moon was now up and some tanks had already appeared in the area of the bridge.

It was now about three o'clock. Dawn would soon be breaking. Away to the left the light of a burning vehicle betokened that Number 3 Company were about, somewhere, but clearly not on their true objective.

In fact that company had found themselves held up on their objective, decided to outflank it and landed themselves among innumerable farms, all of which looked alike. Then the Germans took a hand and it became very clear that the whole ridge was alive with them. So leaving the enemy to go on shooting at each other, Captain Kennedy decided to try the route Number 1 Company had taken. After various vicissitudes they struck their original start-line again. They left the prisoners they had collected and set off again. There was now only an hour left to dawn.

The position was getting desperate, for the forward companies were still out of contact and the few tanks there were had been promised to the Scots Guards who were in an even more precarious position. But there was one stroke of luck for the 1st Battalion Irish Guards that morning. Just as the first tanks passed Battalion Headquarters on their way to the Scots Guards, the wireless operator at Battalion Headquarters turned to the Commanding Officer. 'Number 2 Company is back on the air, sir.' No one knows to this day how Lance-Corporal Holwell, the Royal Corps of Signals operator attached to the company, managed to get his set working under the direct fire of the Germans, but he undoubtedly saved the two companies. There was no time to congratulate him as he was killed seconds later and his set totally destroyed, but in those vital seconds the message to withdraw was passed. Numbers 1 and 2 Companies were on their way. The wounded were put under the cover of the railway bridge and Lance-Corporal Moriarty, the medical orderly, who had been quietly collecting them during the night, elected to stay with them. Then the companies set about fighting their way back.

It was a bloody return. The lines of the railway turned out to be pitted with trenches full of Germans, now very wide awake, and as soon as smoke came down from the guns to help the withdrawing troops it was swept away by the stiff morning breeze. The companies split into platoons and tried to find their own routes. Two of the platoon commanders were killed in the first fifty yards, but the remnants struggled on. Indiscriminate German mortar fire took toll of Germans and Guardsmen alike. Over the bodies of their friends who had been killed the night before, the Guardsmen attacked post after post, driving the Germans before them into the waiting arms and muzzles of the Irish Guards' remaining company, Number 4, who were acting as long-stop. Slightly less than half the two companies who had set out the previous night reached the haven of Number 4 Company that morning. The remnants reorganized into one company behind Dung Farm and prepared for more trouble.

Meanwhile Kennedy and his column were fighting their way from house to house up the left road. Around noon Kennedy came back to Dung Farm with fifty-five prisoners and the news that his company were now in the farm which had been their original objective the previous night. Ordered to continue, they carried on until they met the imperturbable Moriarty and his collection of wounded men under the railway bridge and evacuated them for him.

Another brigade were passed through, and before darkness settled once more had managed to fight their way to a point just short of the Campoleone railway line. Which was as far as anyone was to get for many a day.

Night closed in and the weary battalion settled down to wait for daylight and yet more of the shelling which had continued without cease since the previous dawn. During the night a reinforcement company were sent forward from the beachhead and they could once more muster four rifle companies – but all under strength.

The next two days were quiet as others tried unsuccessfully to get on, suffering desperately heavy casualties in the process. Quiet, that is, except for the incessant, nerve-racking shelling from dawn to dusk. Colonel Scott took the opportunity to send for his 'second eleven', and most commanders and twenty men from each company were sent back to Anzio, where they ate a gargantuan meal, bathed in petrol drums and then slept.

The position which the Allies were now holding on the northern sector of the beachhead was described by an officer as being like the lead in a pencil. Stretching north towards the Alban Hills was this one narrow salient. Four miles long, it was only three miles wide at its widest point, and for the most part a great deal narrower. The Germans had mustered five new divisions, including one armoured, on this single flank. The salient, or the 'thumb' as the Americans called it, was the obvious target.

'Strong counter-attacks to be expected' was how Divisional Headquarters heralded the morning of February 3rd, and with this doleful warning came the rain – a steady, sheeting downpour. Air support was now merely a distant memory, and under cover of the rain who could know what the Germans were preparing? The far too widely scattered companies of the Irish Guards knew they were 'for it', as did those of the 1st Gordons on the other side of the 'thumb'. At 11 PM that night they were proved right.

The attack started with a fierce bombardment, concentrating on the hapless Number 3 Company on the left. Then came the evening infantry; they swept on in mass, very hard, thick and fast after the barrage. The machine-gunners mowed them down, but they came on just the same. Guardsmen Flanagan, Nicholson and Molony, who manned one of the forward Vickers machine-guns, fired all their ammunition, 8,000 rounds, straight into the wavering line of silhouettes, but nothing seemed to stop them; they came on shouting and gesticulating wildly as if doped. As soon as the Vickers had fired its last round three Germans jumped into the trench, shouting 'Hands up, Englishmen!' The three Micks were not standing for any insults; they set about their would-be captors with their fists and made off to Number 4 Company. Meanwhile, the Germans surged in waves against the farmhouse. The enemy operation order for this attack, found later in Rome, showed that two battalions and two special companies had been turned on to Number 3 Company alone. Their casualties must have been staggering, for the platoons cut them down in swathes as they came out of the dark into the glow of some burning haystacks, but the German commanders had their reserves and were determined to beat down the resistance. They shovelled in fresh troops, passing them round the left flank of the company and driving in on the left-hand platoon and Company Headquarters from the flank and rear. They obliterated that platoon and swarmed into the farmyard. 'I never saw so many people killed around me before in all my life,' said Lance-Corporal Donnan, who was captured, escaped, recaptured and escaped again. Captain McInerny, the acting company commander in Kennedy's absence, reported that his headquarters was in a critical position, for 'the Germans are at the door'. And with that last message Number 3 Company went off the air.

Having dealt with the battalion's left flank the Germans now turned their attention to Number 4 Company at the left rear, but with the help of some tank-busters they were beaten off. The plight of the forward positions, where were the other two rifle companies and Battalion Headquarters, was less happy. Cut off and harried from all sides the only answer was to withdraw.

Battalion Headquarters mustered all who were unwounded and set off back down the way they had originally come. The Official History describes their adventures: 'Major Gordon-Watson chose his servant, Guardsman O'Shea, DCM, Lance-Corporal Cross, the orderly-room clerk, and Lance-Corporal Dodd of the Intelligence Section and at their head plunged into the network of ditches and gullies. The rest of the Battalion Headquarters threaded their way behind it in single file. The gully was narrow, overgrown and extremely muddy. The leading section, lightly armed and lightly laden, travelled fast, leaving an ever-increasing gap between themselves and the main body, stumbling along. As they rounded a corner they came face to face with four Germans sitting in a culvert. One of them levelled his Schmeisser sub-machine gun, but dropped it hastily when he was shot in the arm. Some fifteen yards farther on there was another culvert, which contained a further five Germans. This time a blast from the Tommy-guns of Lance-Corporal Cross and Guardsman O'Shea cleared the way, and they stepped over the Germans on to a cattle track.' The whole party ran down another gully, jumped over a bank and landed on top of ten Germans in two slit-trenches. 'I didn't know which way my hat was on,' said O'Shea. The first to recover was Lance-Corporal Cross, though his hands were more accustomed to a typewriter than to fire-arms; the execution he dealt was fearful. The surviving Germans surrendered. It was only now that Major Gordon-Watson realized that he and his little party were alone. There was no sign of the rest of Battalion Headquarters, and no possibility of turning back, so they went on to the main road. Lance-Corporal Dodd 'borrowed' a motor-cycle from a dispatch rider taking cover in a ditch and then, with Major Gordon-Watson on the pillion, drove down the road to the remains of Number 4 Company. Meanwhile, the still imperturbable Cross lined up the twelve prisoners they had managed to collect and marched them down the road, in step and in order of height.

The rest of Battalion Headquarters were now in German hands, for as they plodded through the winding gullies their enemy closed in on them. Rows of grinning German faces, and rows of Spandaus, appeared over either side. The gist of their remarks was apparent to even the non-German speakers, and the column halted with ill grace and a lot of muttering. The Germans slid down the banks, removed the weapons and reversed the order of march. With a guard on each man, the column marched back the way they had come. Captain S. H. Combe, at the head of the column, spoke to his particular guard in a polite voice: 'I am going to do you in when I get a chance.' He did not have long to wait. They emerged from the gully into the sunken road, climbed the bank and continued northwards. As they came into the open a hail of bullets halted and disconcerted the Germans, but not Captain Combe. He walked on, picked up a rifle from the bank and shot his guard. The other Germans did not seem to notice this interlude so, much encouraged, Captain Combe changed the rifle for a Tommy-gun and emptied his magazine into the nearest clump of Germans. Lieutenant J. Bell followed his example by clubbing down his attendant German, and Sergeant McConnell and Guardsman Montgomery snatched the machine pistols from theirs. This was a signal for a general slaughter of the escort. Every Irish Guardsman picked up something – a rifle, a spade, or even a petrol tin – and laid about him. Of the thirty Germans who formed the escort, twenty were killed, nine taken prisoner and one was officially recorded as 'missing'. The Guardsmen dusted themselves down, re-equipped themselves from the general litter of arms, and turned back once more to the railway bridge. And as the day wore on more parties of men dribbled back to Dung Farm from the direction of the forward companies.

By this time the Commanding Officer had arrived. 'B' Echelon, where the 'first eleven' was resting, had received no inkling of anything untoward happening until the information came from Brigade Headquarters that the 1st Battalion Irish Guards were surrounded; one company destroyed, remainder holding on, but isolated. It was a shattering piece of news. The Commanding Officer and Adjutant drove straight to Dung Farm and there the battalion, or those left of it, reorganized to cover the advance of a battalion of a reserve brigade in an attempt to break forward to the completely isolated 3rd Brigade at the point of the 'pencil'.

That night the battalion withdrew to the 'caves' – a collection of old gravel scrapes to the left of the railway. Here they recuperated, and here the survivors of those who had been cut off 'homed' over the long, wet night. At dawn on February 5th they could muster 270 men; a further eighty reinforcements came up that day and so once more three rifle companies, each about seventy strong, could be formed

– but it wasn't much, and too many key and irreplaceable NCOs had gone.

The 'caves' were a grand haven. Although surrounded by a wilderness of shell-pocked, sticky black mud, at least in the gloomy interiors the Guardsmen could sleep and clean their weapons, safe from the almost continuous artillery and mortar fire. At nightfall the companies moved out into their prepared slit-trenches ready to beat off any attack that might disturb their peace. At daybreak they returned to their troglodyte existence, leaving only the Bren-gunners to give protection.

The long-awaited German attack exploded at 0600 hours on February 16th. As far as the eye could range from the slopes of the Alban Hills to the extreme eastern flank, gun flashes spoke of the great offensive. And the eruption of smoke and pulverized earth around the Factory showed where the shells were going.

Then, accompanied by tanks, came the German infantry. Every gun on the beachhead was ranged on this narrow front, and some of the heavies were backed into the sea itself in an endeavour to drop their minimum range so as to reach this new target. Again and again, suffering frightful casualties in the process, the German infantry advancing across open ground were driven back. Again and again they came on. Despite some of the bitterest resistance of the whole Anzio campaign, by the end of the day an ominous gap had been punched in the Allied line.

The battalions of 24th Guards Brigade were called forward to take up a last-ditch defensive position barring the exit of the main road, but the heroic resistance, particularly of the 1st Loyals and the 2nd Battalion of the US 157th Regimental Combat Team, held off the Germans – though only just. Crisis was reached during the long night of February 18th. Attack after attack was mounted by the persistent Germans, and they all but got through. But then even they had had enough, and as dawn broke the onslaught simmered and died down. In retrospect it is now possible to see that this was the vital day and the beachhead was now safe, but no one present on that glorious spring morning appreciated it at the time.

To the 1st Battalion Irish Guards the battle sounded disastrous. All the well-known place-names fell one by one. Hourly the Guardsmen expected to be called – but no summons came until after lunch on the 21st.

The gullies in which the battalion found themselves were in shape roughly like a boot. They were on the average about twenty feet deep, precipitous and covered with brambles. Their slippery sides made it almost impossible for anyone once in a gully to get out again. The top of the boot was to the north and the whole ran parallel to the main road and about 300 yards from it. At the ankle of the boot was a junction. Here was Number 2 Company's position, and an uninviting place it was.

Three days and four nights of bitter close-quarter fighting followed for the battalion. Each German attack was immediately succeeded by an even more ferocious counter-attack. This form of aggressive defence, which the Micks had practised to such excellent effect at 'The Bou', again stood them in good stead; costly as these forays were they maintained morale in a most wonderful fashion and made the Germans keep a respectful distance.

In Number 2 Company's area Sergeant Gundel took note of the position of enemy snipers and then at night he sallied forth and wiped them out; he preferred to work alone. If a shot whistled past a man's head, he was sure to raise a laugh by shouting. 'Take his name, Sergeant Gundel.' A small thing, perhaps, but small things matter in that kind of fighting.

For their part Number 3 Company showed an equal reluctance to remain on the defensive. They turned the gullies into a murderous maze in which men like Major Kennedy, MC,* Lance-Corporal Murphy, MM*, Guardsman Montgomery, DCM,* and Guardsman Adamson, DCM,* hunted down Germans. Montgomery and Adamson were old soldiers and old friends, both Bren-gunners, and two of the

* Awarded later.

few survivors of Number 1 Company then serving with Number 3. Major Kennedy allotted them an outpost position. There they established themselves well forward of the company and for the rest of the day conducted a private war. They could be seen quite clearly as they moved from place to place, covering each other till they came to the final phase of each little operation, when, under protecting fire from Montgomery, Adamson closed in on the sniper or Spandau post and, not having the means or the inclination to take prisoners, shot the Germans with a Tommy-gun. In the afternoon Montgomery was wounded but Adamson continued operations alone for the last three hours of daylight. At dusk he came into Company Headquarters and asked a stretcher-bearer to dress what he called a slight wound on the side of his face. Before this could be done, he fainted. No one knew exactly when Adamson had been wounded, but it is certain that, for some time, he must have been fighting in a state of pain which would have incapacitated a man of lesser determination.

The story of the next few days is therefore the record of the courage and skill of individual officers and men, each continuously fighting his own personal battle, knowing it to be essential to the beachhead. The driving force behind the battalion and the will that knit all these isolated actions into one battle came from Colonel Scott, who kept a firm hand on everybody. It looked so easy the way he did it, no fuss, no hurry, no gestures, just an endless round of visits and an impressive air of absolute confidence in the man he was talking to. Almost as essential to the battalion was the support of the 19th Field Regiment, RA. These remarkable gunners dropped shells wherever the battalion asked for them; they supported section attacks, they broke up large and small German assaults, they put defensive fire in front of ever-changing positions, and on several occasions they laid smoke to cover the withdrawal of one man.

It was a bitter war; as the two sides roamed the gullies, attack after attack was beaten off. When one leader died another stepped into his place. Guardsman O'Brien, created a sergeant on the spot, commanded one platoon with distinction; Guardsman McCracken proved an outstanding section commander. But primarily it was an individual war. Whenever a particularly difficult job required doing, some Guardsman, unbidden, stepped forward and offered to go. There was never any panic when a man or a section became cut off, as inevitably happened. They lay up until dark and then made their own way back to the battalion. The battalions of 24th Guards Brigade were in the enviable position of having only Guardsmen, Depot-trained, in their ranks. As Guardsmen they behaved and in some peculiar way the Depot basic training, considered initiative-sapping by some outsiders, proved the perfect grounding for individual prowess, bravery and enterprise.

As soon as it was light enough to see on their first day in the gullies – February 22nd – the two companies dug deeper into their holes, and there they waited for the next twist of fortune. Number 3 Company spent a peaceful morning, but they were expecting the worst, and at three o'clock it came. Under cover of some tanks in the farm across the main ditch a solid wall of German infantry rose and advanced. Major Kennedy reported cheerfully that he was 'more or less surrounded'.

Then along the track leading to Company Headquarters appeared a full German company. They were within a hundred yards of the slit-trenches. 'SOS right on top of us,' shouted Kennedy over the wireless to the guns. 'The guns responded more accurately than I had dreamed possible,' recorded Major Kennedy later, 'and I called for three minutes' rapid and two minutes' smoke. It would be hard to say what happened during that time, as I only caught glimpses of the Germans, but the concentration was dead on them and about fifty yards in front of us. The smoke thickened and we attacked to round up those Germans who had not run or were dead. The battle was hardly over when we saw about a company of Germans pouring out of the farm ahead and coming down the green grass towards us. It was on that green grass that most of them died, for they were caught flat in the fire of Number 15 Platoon's Bren-guns under Corporal Doran, and by Abbot and Nicholson's Browning.'

Number 3 Company had been successful, but their losses were appalling. In twenty-four hours they had lost seventy men out of ninety. The wounded were a problem until Guardsman Ryan, who had heard the Adjutant assure Kennedy that he would get them out somehow, stepped forward: 'I'll go and get the boys, sir.' Two stretchers were clamped across the back of the jeep and a bandage tied to a windscreen to flutter as a pennant in the wind. Ryan carefully lit a cigarette, reversed the jeep out of its ditch, and slammed it up the track towards 'White Cow' farm. Nobody really expected to see him again, but he reappeared, driving slowly this time, and waited only long enough for the two silent men to be lifted off the stretchers. This was Ryan's occupation for the next two days. He ran a one-man ferry service back to the RAP and then, since there was no ambulance available, from the RAP to the Advance Dressing Station. There are many men walking around Dublin, London, Liverpool and Belfast today who owe their lives entirely to Guardsman Ryan; if they had had to lie out until nightfall they would have died. Ryan's conduct was also an example to the Germans, and they, too, soon set about evacuating their own wounded.

The agony wore on for yet another night, but they knew the Germans would not rest for long. They heard that a battalion of Queen's was due to relieve them before dawn, and Colonel Scott made a plan with Major Kennedy that the isolated Number 3 Company was to pull back if threatened.

The expected attack came at 2 AM. Number 2 Company took the brunt this time and beat them off handsomely; then it was Number 3's turn. It was hopeless to try to hold up the Germans with the few men left in his company, so Major Kennedy decided to pull back. Two officers and seventeen men found their way back to Battalion Headquarters – all that was left of Number 3 Company of the 1st Battalion Irish Guards.

The danger was nearly past. By dawn the rear and left flank were secured by, respectively, a company of Grenadiers and the machine-gun platoon of the Middlesex Regiment. Two fairly fresh companies of the Queen's were now occupying Number 3 Company's old position.

They heard that the Duke of Wellington's Regiment was to relieve them that night. As soon as it was dark Battalion Headquarters trailed across to meet the new arrivals, and as the hand-over was in full swing the Germans struck a final blow as they banged down a vicious concentration of 88-mm shells and mortars. The Duke's suffered severely and the Irish Guards lost Sergeant Wylie, DCM – 'solid Jimmy Wylie, as brave as a lion and as patient and gentle as an ox, who had survived the worst of three campaigns and was looking forward to a fourth' – and then Corporal Moriarty, DCM, hit by a fragment of a mortar shell as he went to help his old friend.

Lieutenant-Colonel B. W. Webb-Carter, commanding the Duke's, wrote: 'As the tall Guardsmen filed out, leaving us the heritage of death and desolation they had borne so long, a peculiar sense of isolation struck us. In all the long-drawn-out crucifixion of the beachhead, no positions saw such sublime self-sacrifice and such hideous slaughter as were perpetrated in the overgrown foliage that sprouted in the deep gullies.'

The battalion marched back as though in a dream. In the rear areas they lay where they dropped. At dawn the cooks brought mugs of boiling tea and dixies of bacon, but the men were still too tired to eat. In the evening, after a day asleep in the dugout and an hour cleaning weapons, they remembered they were hungry. The next day they remembered they were Guardsmen, and if they did not, they were reminded by Drill-Sergeant Kenny.

For ten more days they remained in the beachhead. The Divisional Commander, General Penney, came to say goodbye. The General told them, 'I can't let you go without saying how much I shall miss the Regiment, not only collectively, but individually. It has been a privilege and a sadness to have had your battalion in the 1st Division during these days, but their achievements, their unfailing response and their willing fighting spirit have been an inspiration. I regret your departure enormously, especially as I fear there is little chance of your rejoining us, but I am glad of the opportunity you will have of refitting and reforming. I wish you all well, and I shall never forget what you have done and the sacrifices you have made.'

On March 7th 267 all ranks of the 1st Battalion Irish Guards sailed from Anzio. Including reinforcements, 1,080 had landed.

St Patrick's Day came and went. For once no one seemed to notice it – they were still mentally recovering from the punishment they had received at Anzio. Rehabilitation and hospitality at the hands of the Italians helped a lot, but there comes a time when men have had enough and when they are no longer human beings but trained zombie-like fighters – a good few of the survivors of the 1st Battalion were in this case.

Earlier Colonel Scott had written to Regimental Headquarters that there were no more reinforcements available in the theatre and, whereas the other battalions in 24th Guards Brigade had battalions of their own regiments fighting elsewhere in Italy, the Irish Guards were alone. Thus it was clear that 1st Battalion Irish Guards must be due for relief and return to England.

But before they went, Rome was captured, and through the good offices of Monsignor O'Flaherty, who had masterminded the escape pipe-line in Italy, the Pope agreed to receive a company of the battalion in special audience. It was the most memorable occasion of their lives for those lucky enough to be present, and as they kneeled to receive His Holiness's Blessing the emotion of the moment was too much for one Mick. Rising to his feet, he said in a strangled voice, 'May God bless you, Holy Father' – to which Pope Pius quietly replied, 'I hope He will, my son.'

Chapter 12

North-West Europe: Normandy to the Seine

For those at home the long wait was now nearly over. In April 1944 the Guards Armoured Division under command of Major-General A. H. S. Adair, DSO, MC, left Yorkshire; of the two brigades, 5th Guards with the 2nd Armoured Battalion Irish Guards went to Brighton, and 32nd Guards with the 3rd Lorried Infantry Battalion Irish Guards went to Eastbourne. There they had a grandstand view of the massive assembly of force for the landing on the mainland of Europe; and they watched the Allied airpower roaring overhead on D-Day, June 6th.

ORGANIZATION OF THE GUARDS ARMOURED DIVISION

5th Guards Armoured Brigade	*32nd Guards Brigade*	*Divisional Troops*
1st (Motor) Battalion Grenadier Guards	5th Battalion Coldstream Guards	Armoured Reconnaissance Regiment–2nd (Armoured) Battalion Welsh Guards
2nd (Armoured) Battalion Grenadier Guards	3rd Battalion Irish Guards	
1st (Armoured) Battalion Coldstream Guards	1st Battalion Welsh Guards	
2nd (Armoured) Battalion Irish Guards		

NB Also serving in North-West Europe were the 6th Guards Tank Brigade (later 6th Guards Armoured Brigade). They were an Independent Brigade unconnected with Guards Armoured Division.

Although final preparations were finished long before, the Guards Armoured Division did not land in France until D plus 21.

By July 6th the build-up in Normandy was nearly complete. The great port of Cherbourg had fallen to the Americans at the end of June, and on their front Second British Army had pushed forward about fifteen miles from the coast and were within reach of Caen. The battle of attrition was on in earnest. Under heavy pressure, division after division of the sparse and valuable Panzer reserves were being pulled into the battle. The scene was now set for the break-out.

The country could scarcely be worse for armour. South of Bayeux stretched the Bocage – a land of small fields, of densely planted mature orchards, of deep narrow lanes with thick banks surmounted by strong hedges. Visibility was minimal. The devastation of war was very much in evidence, and all that remained of the once fine herds of Normandy cattle were the stinking, rotting, swollen corpses which littered the fields. But to the east beyond Caen the country opened out into rolling plain and it was to here that the eyes of the planners turned – perhaps mesmerized by its geographic resemblance to the desert, a totally misleading comparison.

The overall plan was for the British and Canadians to hold and enlarge the east of the bridgehead while the break-out was effected by the Americans towards St Lo and Avranches. The principal aim was to attract those German reserves that still remained, particularly the Panzer Divisions, away from the US front, for Allied tank losses could be replaced – those of the Germans could not. This was the concept behind Operation Goodwood.

To the east of Caen lies the broad plain of La Batte de Hogue, which rises to a point north of Cagny and then descends in a gentle, unbroken slope to the village itself on the outskirts of which the country abruptly changes to close bocage once again. It was here that the Guards Armoured Division were to have their first battle.

Leading the advance of 5th Guards Brigade, the 2nd Armoured Battalion Grenadiers breasted the rise north of Cagny and immediately lost half a squadron of tanks.

From now on the enemy defence was bitter. Gradually, as the day wore on, the tanks of the 2nd Grenadiers and 1st Coldstream fought their way into Cagny, but any progress

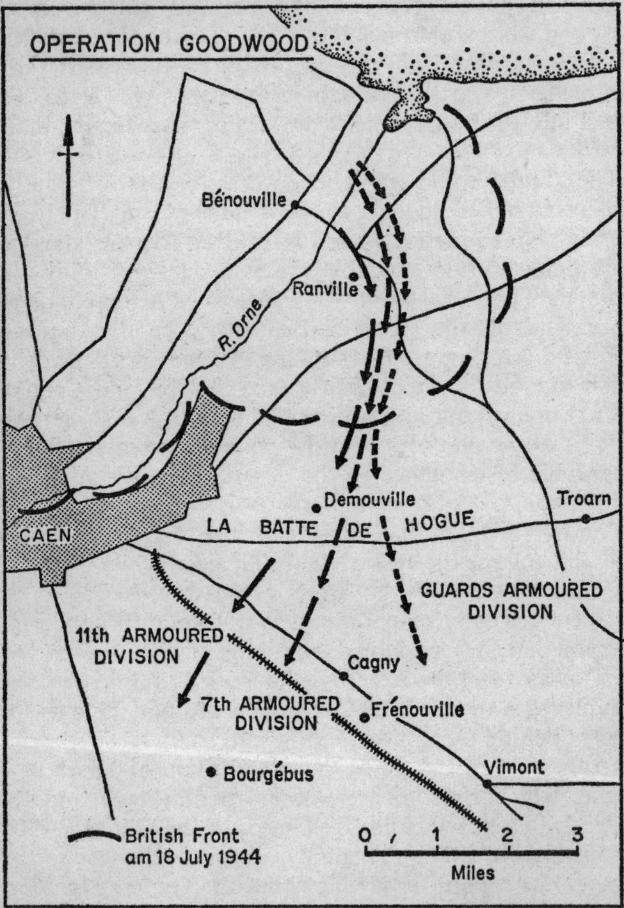

beyond the railway line to the south of the town was bitterly contested.

At five o'clock the 2nd Irish Guards were ordered to try

the left flank. Instructions to 'follow the pylons' were passed down from tank to tank. The squadrons formed up in line ahead. 'There's too much allez-allez about the whole thing,' was the comment of one officer, and he was about right.

All went well at first, but one of the troop leaders of the forward squadrons, Lieutenant J. Gorman, drove into a bog and by the time he had switched tanks he had lost touch with the others. He continued alone and as he came to the brow of a small hill he let out a wild cry, 'Gunner!' for there, not more than two hundred yards away, were four enemy tanks: a Royal Tiger – the first seen on the Western Front as it happened – an ordinary Tiger, a Panther and an old Mark IV. The Germans were equally surprised and were luckily all facing the wrong way. But no shot came from the Sherman: 'Guns jammed, sir' – Guardsman Schole's voice was despairing. 'Oh, Christmas, why?' By now Gorman saw what he must do. 'Driver, ram!' he shouted. Lance-Corporal Baron let in the clutch and 'Ballyragget' crashed through the hedge and careered down the slope towards the enemy. The tank slid down beside the long barrel of the 88-mm gun and struck the Royal Tiger at the rear of its right track – totally immobilizing it. Both crews escaped.

Once launched there was no controlling the battle. The 'air' was bedlam – everyone seemed to be reporting 'Hornets' everywhere (the code name for enemy tanks); this was their first taste of the 'fog' of war.

However, it soon became apparent that a strong screen of enemy tanks and anti-tank guns was well sited about a mile east of Cagny – so it was decided to try to outflank them. But the country on that side was broken, the German SS defenders skilful and stubborn, and night was falling fast. 'No further advance' was ordered at 9.30 PM. The remaining tanks of the 2nd Battalion pulled back to harbour – or zareba as they insisted on calling it – on the outskirts of Cagny. It was now the turn of the infantry.

The 3rd Battalion Irish Guards had spent an unhappy day moving in easy stages in line ahead behind 5th Guards

Brigade, feeling very vulnerable in their unarmoured transport. They were lucky and weren't shelled, until, when passing through the minefield, they lost a company commander fatally wounded by a splinter. At last they debussed west of Cagny and started to dig in. As soon as the slit-trenches were of a respectable depth they were off again. It was by now pitch dark so, setting a compass bearing, the battalion moved forward to Frénouville and beyond.

No sooner were they under way than the orders were changed: Cagny itself was now the objective. It was too late to retrieve Major A. R. Eardley-Wilmot with Number 2 Company, who had set out at a lively pace and were not seen or heard of again until the next morning, and half of Number 1 Company which had followed them and could not be caught. So Colonel Vandeleur summoned his remaining company commanders and began to describe the situation. 'Never,' he said, 'was there more confusion at the start of an operation...' and with that the tank on which he had inadvisedly propped his map drove off. After a fine chase the map was retrieved and he finished his orders.

Marching by compass, Colonel Vandeleur led the 3rd Battalion Irish Guards to battle. When they reached Cagny it was to find 'bomb craters as big as buses, and stone-walls surrounding the village like an old-fashioned fort, burst drains, the stink of dead cattle and dead Germans everywhere'. To further enliven the proceedings some German fighters swept down in the moonlight with their cannons blazing. 'X' Company, the attached Scots Guards Company, were at first held up in the ruins, but then, having sorted themselves out, the whole battalion marched forward again in single file. 'We wound our way through standing corn beyond the village for more than three hundred yards when the crack of Spandau and flash of small-arms appeared straight ahead,' described Colonel Vandeleur. 'Very lights and star-shells shot into the sky and every form of unpleasantness opened up on us, including those wicked old beasts the "Moaning Minnies". To make things worse it was painfully obvious that our own troops, including some enthusiastic two-inch mortar crews, were also shooting at us. However, luck came our way. We had inadvertently moved through a screen of our own 2nd Battalion, who gave us some sound advice. It is not easy to cast a plan in the dark, so I spun a coin instead – heads, move forward in Wellington style, bayonet to bayonet; tails, consolidate where we were. Luckily it came down tails so we hatched out some form of defensive layout and dug like beavers.'

That night the Headquarters of the two Irish Guards Battalions were together among the tombstones of Cagny cemetery. There was still no news of the missing Number 2 Company until, the next morning, the Company Commander suddenly emerged from a ditch to say that he was sharing a hedgerow with a few Germans short of Frénouville some fifteen hundred yards away.

That company's navigation had been excellent on the previous night. Approaching the village from the north, they had seen no sign of life, so while they lay flat near the roadside, a section was sent forward to reconnoitre. As Sergeant Cooke and his seven-man section reached the outskirts of the village a Spandau opened up. Emptying his Sten in the direction of the flashes, Sergeant Cooke rushed the post himself, caught the German gunner by the throat and throttled the life out of him, being badly bitten in the process. The bite later went septic and Sergeant Cooke had to be evacuated.

Then the Germans woke up. Spandaus opened up on the company from Frénouville itself, and it was only too clear that they had walked into the middle of a very strong position. It was impossible to do anything but dig – and they dug for dear life, fully aware that anyone above ground come the dawn was a dead man. Conscious that only he could lead back his company, as soon as it was light enough to see, Major Eardley-Wilmot crawled off to arrange for their extrication. He was recommended for the immediate award of the Military Cross – the first decoration earned by the 3rd Battalion Irish Guards – but was killed before it was awarded.

So Operation Goodwood came to a halt. No headway could be made against the determined German opposition, and on the afternoon of the 20th the weather broke. The place was a quagmire within half an hour, and further operations on the Cagny front were soon an absurdity. To add to their discomfort the mosquitoes came out in force; so, shelled and mortared, soaking wet and badly bitten, the Guards Armoured Division retraced their steps.

From the point of view of the 3rd Battalion, 'The one lesson we learnt at Cagny,' wrote the Commanding Officer, 'was, needless to say, the old one of the futility of a night advance over unreconnoitred ground.' From a broader perspective, the Germans had been forced to throw into battle yet more of their precious armour, leaving a bare two Panzer divisions opposite the American sector. The way was now clear for the great advance of the United States First Army.

On July 25th the Americans began the operations which were to result in the break-out from the Normandy beachhead. Following up a fierce bombing attack on the St Lo area by nearly 2,500 aircraft, the Americans hammered on against stout resistance. They were able to advance a couple of miles on the first day; Perriers fell on the next; Avranches, forty miles to the south, was captured on the 30th. The way to Brittany and beyond was open at last.

But General Patton's left flank was dangerously open. It was to discourage the Germans from any idea of interference in that quarter and to keep them occupied on the Second Army front that an attack was mounted in the Caumont area on July 30th, the day that Avranches fell.

The country south of Caumont consists of a series of east–west ridges interspersed with streams, all of it bocage at its best. The resident Hun had not been inconvenienced for several weeks, so prior to the attack the RAF 'softened up' the positions, with considerable success. The initial attack was mounted by 11th Armoured Division with the independent 6th Guards Tank Brigade in support. By noon the forward defences had been over-run but the momentum was slowing. By a fine piece of work the 4th Tank Battalion of the Coldstream in 6th Guards Tank Brigade – whose Churchill tanks could cope with the bocage banks – secured Point 309, the most north-westerly point of Mont Pinçon Massif. At this juncture the Guards Armoured Division were ordered to pass through and take up the advance.

One lesson learnt at Cagny was the inadvisability of committing tanks without infantry in the oppressively close country of the bocage. So the infantry/tank battalion groups in the Guards Armoured Division were born. Initially it was a matter of temporary convenience that in 5th Guards Brigade the 2nd Armoured Battalion of the Irish Guards were grouped with the 5th Battalion of the Coldstream – who happened to be sitting in the next field. Similarly in 32nd Guards Brigade the 1st Armoured Battalion of the Coldstream were grouped with the 3rd Battalion Irish Guards; and the 1st Motor Battalion and 2nd Armoured Battalion of the Grenadiers joined forces. Only the Welsh Guards remained unaffected, one battalion in Cromwell tanks keeping their reconnaissance role and the other staying as an independent infantry battalion. Ultimately the Guards Armoured Division were organized into exclusively regimental groups, and there is no doubt that this new formation had much to do with the outstanding success of the division during the rest of the campaign in North-West Europe.

The intention to make this reorganization was clear enough. But unfortunately, the participants had not been told. The first the 2nd Battalion Irish Guards heard of the new idea was when the divisional commander ordered them and 5th Coldstream Battalion 'to advance together as a battle group'. As the Official History records: 'It took the two colonels and three hundred officers and non-commissioned officers an hour and a half to rearrange the two battalions. Out of the scrambling confusion of men, tanks and trucks, guns, carriers and motor-cycles, shouting officers, desperate sergeants and patient Guardsmen, there emerged the Irish-Coldstream Group, neatly formed up along the main road. Number 2 Squadron Irish Guards was leading with Number 4 Company Coldstream Guards, then came

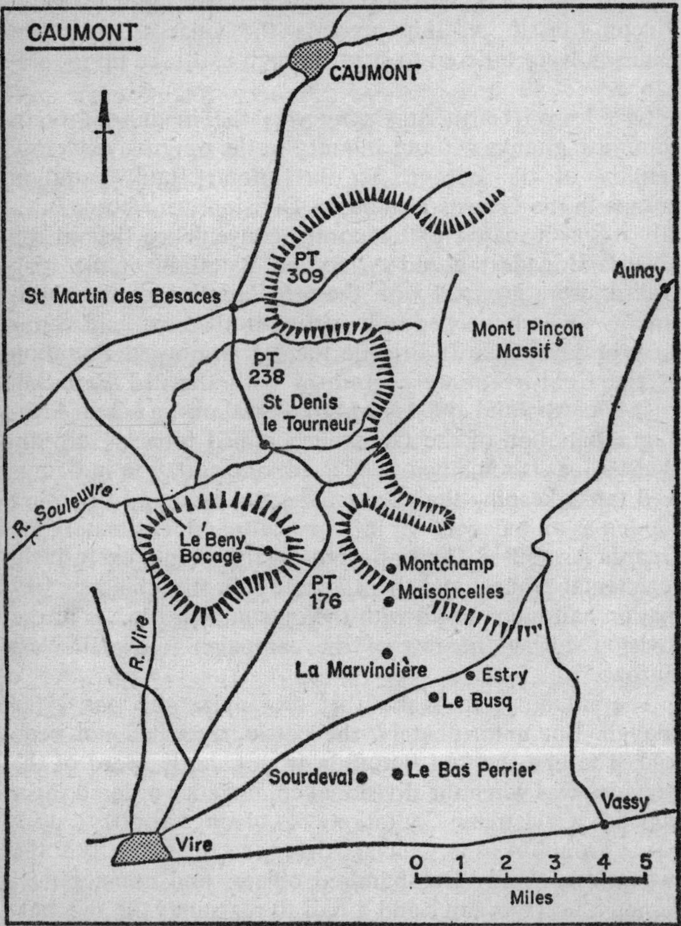

two infantry companies, followed by Number 3 and Number 1 Squadrons, and finally the reserve Coldstream company.' Their main objective for the night was a long, high ridge six miles south of St Martin des Besaces and the collection of houses scattered along the top of it called Le Beny Bocage.

The newly formed group moved forward into a welter of transport along the one serviceable road in the area – it included support for the 15th Scottish Division and 6th Guards Tank Brigade, 11th Armoured Division's 'tail', various corps units as well as their own – and the traffic problem became 'grotesque' as described by Lieutenant-Colonel Finlay, commanding the 2nd Battalion. The inevitable delay caused by the traffic turned out to be crucial.

Ordered to press on to Le Beny Bocage and harbour at Point 238 for the night, the Irish/Coldstream Group were no sooner launched than the Germans opened up every weapon they had on the advancing troops. The squadrons scattered like 'a covey of partridges'. Fording a stream at the bottom of the slope they stormed up the opposite hill. Visibility in the high corn and amidst the hedges in full leaf was virtually nil. Added to which it was nearly dark. So the group called a halt and dug in for the night. Neither side was aware that the other was only some two hundred yards away over the crest.

The following morning the advance continued. They occupied Point 238 without much trouble, losing one tank on the way. It was that of Pipe-Major Crozier; hit from the front, out of control, and blazing furiously it started to roll backwards down the slope. At the bottom of the hill and directly in its path lay the Regimental Aid Post still full of wounded men who could not be evacuated the previous night. They saw the monster coming towards them, gathering speed as it came. Only a bank, thin by bocage standards, intervened. 'We prayed as we have never prayed before,' said one man. The tank hit the bank below a white-thorn tree, swayed and stopped. Pipe-Major Crozier was badly burnt and his thigh was smashed. The gunner, Guardsman Cuthbertson, was also horribly burned and had his foot hanging by a thread. When the stretcher-bearers got to the tank the Pipe-Major was hauling Cuthbertson under cover, and both of them were singing *The Mountains of Mourne*. 'There's the Micks for you,' said the battalion proudly.

From then on the situation became confused. In that close country, control of the battle was impossible. It was a contest of troops and platoons, of individual tanks and infantry sections pitted against anti-tank guns, self-propelled guns and a stubborn, determined, fanatical enemy. The reports on the 'air' sounded as though a menagerie had escaped – Tigers, Panthers, Hornets and Crabs everywhere, some dug in, some moving north, south, east and west. 'We gave up counting the brutes,' said an officer.

For the next forty-eight hours they played hide and seek among the hedges and ditches around Estry. They observed bitterly that 'no matter what is said in Parliament, the German guns penetrate our armour, and our 75-mms do not penetrate theirs.' Artillery support was not forthcoming either, for the battalion's own battery of the Leicestershire Yeomanry had chosen to harbour one night in the same field as some German Panthers, and lost three of their self-propelled guns and most of their transport in the ensuing mêlée.

The Irish/Coldstream Group, by then joined by a squadron of Welsh Guards, were at the point of a narrow salient. To the right the Germans were in force at Le Busq, while to the left they could see streams of enemy armour going forward towards Montchamp, then in the process of being cleared by the 3rd Irish.

On August 4th the German advance ceased abruptly and they started retreating into the jaws of the group, which duly exacted toll. But their own position was highly precarious, for the sole supply route was under continuous artillery and mortar fire, and on one stretch under direct tank observation. An ambulance convoy was formed by the simple expedient of unloading some lorries, emblazoning them with Red Crosses and taking them back slowly and deliberately. The Germans left them unmolested. A convoy of another sort was forming at the other end. 'A' Echelon was collected together by the Transport Officer, Captain the Lord Claud Hamilton, and the drivers ordered to run the gauntlet one by one. Cheerfully and driving at full speed they got through – they were far too fast for the sluggish traverse of the German tank guns.

They stayed around Le Busq for four more days. Rumours of another push came and went; meanwhile the 2nd Armoured Battalion suffered. Already they had had seventeen officer casualties – nearly half the officer strength – including two of the three squadron commanders; the third was wounded a few days later. There were no more officer reinforcements available in Normandy, so urgent calls went back to Regimental Headquarters in England. It was about this time, too, that Colonel Finlay handed over command of the 2nd Battalion to his second-in-command Major G. A. M. Vandeleur, a cousin of Colonel J. O. E. Vandeleur of the 3rd Battalion.

The 3rd Battalion had also been having their share of excitement as they struggled to clear the centre-line behind 5th Guards Brigade. They had learnt their lesson at Cagny; now the advance was made to carefully detailed limited objectives, each reconnoitred and each properly planned. As a result their progress, although slow, was at least sure. Short of Maisoncelles they paused for the night and dug in. It was as well they did.

Just before dawn the Germans attacked. Skirting the leading companies they ran straight into Battalion Headquarters, who bore the full brunt. To the roar of 'Mow them down!' from Colonel J. O. E. Vandeleur, Battalion Headquarters set about repelling their unwelcome visitors. Many of the prisoners who were brought in afterwards were Tartars, who looked as if they ought to be wearing pigtails. Captain E. Udal spent a long time interrogating one unhappy man with a pick-handle and in German, until the man thought of one intelligible word – 'Russki'. Battalion Headquarters followed the Commanding Officer's advice to the full. He himself was delightedly firing the gun mounted on his scout-car; the Signals Officer, with some clerks and signallers, climbed trees and loosed off every form of weapon at whatever they thought was a German, and others were belting away with German machine-guns which they had collected during their time in Normandy. It was a highly successful little affray.

The next days saw the battalion with their Coldstream tanks working their way forward against still stiff resistance. Then the Germans started withdrawing eastwards and the advance quickened.

On August 9th the Irish/Coldstream Group were ordered to take over the tip of the salient secured by 11th Armoured Division in the area of Sourdeval. A horrible area it turned out to be, and there seemed to be a deathly stillness all around this unpleasant place. Never before had they seen such a shambles of burnt-out vehicles and shelled orchards. It smelt terrible, a new brand of Camembert cheese this time, more ripe than ever. Strong patrols the next day and following night confirmed what they had already suspected – that the town and country beyond was held in considerable force by the 10th SS Panzer Division, whom they had met before at Cagny, and also by some parachutists.

Late on the night of the 10th the 3rd Battalion received orders to attack Sourdeval the following morning with a squadron of tanks and one field battery in support. Protest as he might, even Colonel J. O. E. could get no more assistance, for the battalion's attack was only part of a full divisional effort designed to cut the main road from Vassy to Vire, which was one of the few west–east escape routes left to the Germans in Normandy.

As the Official History records, 'The battalion attacked at nine o'clock on the morning of August 11th. The vista in front of them was appalling. From the top of the Sourdeval ridge the open fields of ripe corn sloped down to the stream in the bottom of the valley. Beyond the stream the ground rose again steadily for five hundred yards up the line of the main road. A man on this rise could see a mouse move on the forward slope opposite him. On the left flank the Germans entrenched on the high ground could sweep the cornfields with enfilade fire. The battalion, therefore, had to cross the crest of the ridge, walk down the forward slope with no protecting support straight into the faces of the paratroopers and their Spandaus, completely under the view of German observers and the fire of their tanks and guns on the high ground to the front and flank.'

It was a tragic advance; the moment the leading companies crossed the crest of the ridge they came under devastating fire, the ground being scythed by the Spandaus and heavy German artillery. The German guns and tanks on the far slope could not miss the supporting Coldstream tanks as soon as they appeared over the crest, and promptly destroyed them. The companies pressed on. For a long hour they walked straight down the hill, pushing through the corn or plodding through the root fields. Spandaus, firing straight up the lane and its hedges, drove the Guardsmen away from even this slight cover. Strung out in open order, they kept on moving forward.

There being no cover and with fire coming from all sides, they fixed their eyes on the stream ahead of them, ignored the fire, and advanced steadily. Each man talked occasionally to the man five yards on either side of him till one or the other, or both, or he himself, or all three, were hit. The sharp crack of the sniper's rifle punctuated the clatter of machine-guns and the steady thump of mortar bombs. The wounded lay where they fell or, if they could, crawled to the shelter of the standing corn. The burning phosphorus of the mortar bombs fired the dry stalks. The companies choked and coughed their way through a swirl of smoke and flame. The wounded watched the flames creep on to them. Some dragged themselves painfully out of the corn back to the open root fields. The stretcher-bearers shed their equipment and worked frantically to shift the unconscious till they, too, dropped wounded in the corn.

By ten o'clock the two leading companies had reached the stream, leaving thirty-three dead and seventy-two wounded behind them, three-quarters of their combined strength, and all but three of their officers. But now, at long last, there was something solid to get at. All along the line the Guardsmen jumped into the trenches, slaughtered the paratroops with cold rage and pitched their bodies out of the trenches.

The position of the remnants, though, was perilous in the extreme. 'They couldn't go forward, and wouldn't go back.' Out of touch with the rest of the battalion, for their wireless had been destroyed very early in the proceedings, they waited patiently.

To advance further was obviously futile with so little artillery support as Colonel J. O. E. had at his disposal; he determined to call off the attack and extract the survivors of his two companies. Smoke was the only way he could do it, but it would still take some time to collect enough. While they waited Captain A. Drewe, second-in-command of the attached 'X' Company Scots Guards, volunteered to tell the forward companies what was afoot. Taking two companions he set off first to destroy a troublesome machine-gun post on the way. One of his men was killed and the other wounded, and Captain Drewe himself was struck in the leg, but he killed the Germans and struggled forward to shout the plans to the forward companies. Promptly at half past four the smoke came down and Captain Drewe went forward once again to lead in the remnants. The 3rd Battalion Irish Guards was reformed into three companies, and pulled out of the line. The Household Cavalry took over. Sourdeval was a disaster. The only redeeming feature was the conduct of the attacking companies. But there was no doubt that the 3rd Battalion felt very bitter about the whole affair, and well they might.

The next weeks were gloriously idle. Although strictly on compo rations, the 3rd Battalion had found themselves in an area where it was easy to get cheese and butter, vegetables and cider. Here they said farewell to their borrowed Scots Guards company, who were bound for the 1st Battalion Welsh Guards as reinforcement. The final party was a memorable occasion. Instead of the Scotsmen, with whom they had got on so well, they welcomed a number of reinforcements including the complete Number 4 Company from the 1st Battalion. 'All veterans, and it was grand to see them,' Colonel Vandeleur wrote. By the end of the month they were ready and on August 30th crossed the Seine.

The 2nd Armoured Battalion were also recuperating in the comparative peace of the Normandy countryside. They heard with delight on August 19th of the closing of the Falaise Pocket, and the virtual elimination of German resistance west of the Seine. On the 24th Paris fell; on the 27th the 2nd Battalion were told to put their tanks on transporters and move to L'Aigle – nearly seventy miles towards Germany. Only then did it start to dawn on them just how complete was the German débâcle. They, too, crossed the Seine on the 30th. The Guards Armoured Division were now poised for the next phase.

Chapter 13

North-West Europe: to Brussels and beyond the Rhine

At two o'clock the next morning the 2nd Armoured Battalion were off, leaving the 3rd Motor Battalion in reserve. By ten o'clock Beauvais had been reached, thirty miles away. They passed on their way hampered by surrendering Germans and crowds of delighted Frenchmen alike. By nightfall they had crossed the Somme; they had covered sixty miles that day.

11th Armoured Division on their left had meanwhile captured Amiens, and on the next day the 'gallop' continued. The 2nd Battalion Irish Guards reached Douai, fifteen miles north of Amiens, that night to be greeted by the warmest welcome so far. The Resistance as always were helpful, switching signposts – a practice which had interesting repercussions for both sides – looking after prisoners, and providing information. To the battalion's disgust, however, they could not find a German paymaster amongst the assorted captives, for they had heard of the good fortune of the Grenadier Group who had caught one with 600,000 francs on him.

They stayed the night in the town, Battalion Headquarters occupying a fine café in the main square. Meanwhile, it was possible to get through by public telephone to the Mayor of Lille, thirty miles to the north, who obliged them with the information that the Germans were in full retreat, and seemingly had no intention of defending the place.

On the afternoon of September 2nd the two Irish Guards

battalions were combined and the Irish Group was born. The logical successor to the haphazard grouping which had taken place in Normandy, the Guards Armoured Division now comprised four regimental groups: the Grenadier and Coldstream Groups under the wing of 5th Guards Brigade; and the Irish and Welsh under that of 32nd Guards Brigade. From its inception the new reorganization was an outstanding success, and infantry/tank cooperation took on a new meaning. Perfect trust and mutual cooperation were the keystones – the two battalions spoke the same 'language', most of the officers and NCOs and a lot of the Guardsmen knew each other. Each company worked with the same squadron; each platoon with the same troop and each section with the same tank – and they knew that their tank would not let them down. As senior Commanding Officer, Lieutenant-Colonel J. O. E. Vandeleur of the 3rd Battalion took command of the group, while Lieutenant-Colonel G. A. M. Vandeleur remained in command of the 2nd Battalion. Colonel J. O. E.'s first orders to his new group were: 'Enemy Information – one word, chaos. Our Intention: the Irish Group will dine in Brussels tomorrow night!' Brussels was sixty-five miles away.

Few can have slept well, if at all, that night, least of all the Technical Adjutant. Would the tracks, the bogies, the engines last? Crews worked until midnight and beyond on their tanks. Elation was in the air, and a potentially dangerous euphoria was taking command of normally sober and unemotional minds.

September 3rd dawned chilly and damp. At 6 AM a squadron of the Household Cavalry and the Welsh Group passed through to take up the running. Some isolated resistance slowed the advance to start with, but not for long. Then they were off; as the Official History describes: 'The 2nd Battalion had no mechanical trouble; for years they had been preparing their tanks for a day like this. Through the Belgian villages roared "St Patrick", "Ulster", "Leinster" and "Connaught", "Achill", "Bantry", "Cloneen" and sixty-seven other Irish place-names. The rattle of tank tracks on cobbled streets brought the incredulous inhabitants out of their houses to wave and cheer wildly. Contrary to expectations the Belgians were more demonstrative than the French, and more practical; they established plum, apple and pear "points", and lobbed bottles of "Lion d'Or" beer into the vehicles. Some stray American armoured cars appeared from a side road; they cheered the group and the group cheered them.'

Tournai, Ath, Enghien came and went. By six o'clock they were ten miles from Brussels. As light was failing they reached its southern outskirts. While the Welsh Group headed straight for the centre of the city, the Irish swung right-handed to cut off the Germans from the east. Despite the obstruction of huge, delirious crowds, bunches of hydrangeas and a hail of unripe fruit they reached the main gate of the park of the Bois de la Cambre by nine o'clock.

And now they learned what 'liberation' really meant. The Bruxellois went mad. The lights were turned on, shops lit and the streets were a thronging mass of delighted humanity. The tanks and vehicles, already looking like floats at a carnival, were festooned with flowers and coloured streamers. Anyone unwise enough to ride in an open car was smothered with kisses by old and young alike – seasoned veterans of many campaigns wept openly. Everyone was so tired after the last days when they had advanced 180 miles with virtually no sleep, borne on by their nerves and the overwhelming gratitude of those they had liberated, that they could hardly stand, let alone think. They just let the delirium of the happy crowds flow over them.

On September 6th the Irish Group started off again. In the meantime the Grenadier Group had taken Louvain on the 4th and on the afternoon of the 6th the infantry of the Welsh Group managed to force a crossing over the Albert Canal at Beeringen. During the night sappers established a Bailey bridge over the remains of the former road bridge and the whole group crossed at dawn. It was sobering to remove the flowers and traces of 'liberation' and once again get down to work.

The two brigades fought their way north and Helchteren fell to the Irish Group on the night of September 7th, but Hechtel was proving more sticky and the Germans, a

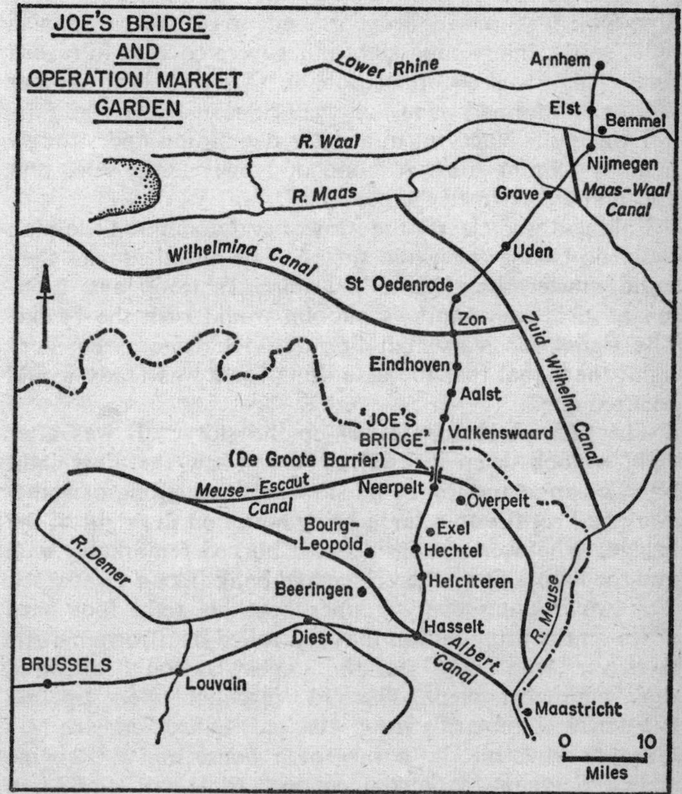

mixture of paratroops and Dutch traitors masquerading as SS, were resisting with fanatical stubbornness. It was therefore decided that the Irish Group should outflank the town from the east and make best speed for Exel, as a preliminary to a drive to the Escaut Canal.

No sooner had they arrived, though, than the 2nd Household Cavalry Regiment, who had been ranging widely, reported a brand-new, unmapped military road which ran straight through to Overpelt and then turned left to the Escaut Canal at De Groote Barrier.

This was too good an opportunity to miss. Number 1 Squadron Group had been ordered 'to go like a Spanish bull', so the intervening eight miles were covered in record time. There was no opposition. In Overpelt itself another scout car informed them that the bridge at De Groote Barrier was still intact but mined for demolition and strongly held by 88-mm guns. A troop of Honey tanks were sent ahead and confirmed the news.

Boldness was clearly the answer and a simple little plan evolved. Under the protecting fire of all the tanks they could muster, Lieutenant S. Lampard's troop and Lieutenant J. Stanley-Clarke's platoon would rush the bridge. The signal for maximum fire would be one green Very light; the signal that the assaulting force was ready would be a red one.

The Official History takes up the story: 'It was after eight o'clock when the waiting crews saw the Very light burst green at the top of its slow arc. Something or somebody had set fire to a large brick house on the right of the bridge. Whatever was inside it, it burned remarkably well and the flames lit up the bridge and banks like a theatre set. For two minutes every gunner kept his right foot hard down on the firing button that controlled his Browning. All co-drivers who could see, and some who could not, fired their main armament. The red Very light went up and Lieutenant Lampard's troop charged. Lance-Sergeant McGurren's tank hit the corner of a house and the engine stalled. Sergeant McGurren swopped shots with an 88-mm while the driver wrestled with the engine. In the second tank Sergeant Steer shouted to his driver to swing past Sergeant McGurren and drive straight for the bridge, where an 88's tractor had now been set on fire by the shooting. The tank crashed through the flames and explosions. Corporal Barrett, Lieutenant Lampard's driver, kept his tank hard on its tail, followed by Sergeant McGurren.'

Colonel J. O. E. was dancing with excitement on the top of a slag-heap. 'It's the drivers who've done it – the men behind the wheel!' The platoon led by Lieutenant Stanley-Clarke and Lance-Sergeant Helliwell ran across behind the tanks, sprayed by Spandau fire, and flung themselves into the ditches. An excited German surrendered, shouting, 'Well done, well done!' They had got the bridge. The rest of the squadron and the company swept over to join them. The first to follow was a sapper officer, Captain Hutton, with a party of four Guardsmen of Number 2 Company – Guardsmen Trimble, Mollard, Colthorpe and Fielding, who had been detailed as 'you, you, you and you'. They climbed over the side of the bridge and cut the cables. The machine-gun bullets were drumming into the wooden beams, but they worked steadily till they had removed the detonators from the charges in the piers. Number 2 Squadron and Number 3 Company went over to reinforce the original group and took up a position in a wood on the left of the road, just over the canal; the position was now secure and the Escaut Canal bridge was ours. By a sad twist of fate the squadron commander, to whom much of this success was due, Major D. Peel, was killed the next day as the Germans tried again and again to recapture the bridge.

The success of this operation had saved the 2nd Army several days and enabled the date of the next advance to be brought forward – so said the Corps Commander, Lieutenant-General B. G. Horrocks, who had come up in person to congratulate the group. And then he gave the Irish the doubtful honour of leading the next advance – Operation Market Garden.

With the bridgehead over the Escaut Canal secured all was ready for the next phase. (See map, p 181.) Across the path of Second British Army lay six major water obstacles – the Wilhelmina, the Zuid Wilhelm and the Maas–Waal Canals, and the Rivers Maas, Waal and Lower Rhine – as well as innumerable smaller streams. If it was possible to capture intact the bridges across these waterways, thus cutting off the quarter-million Germans in Holland, there was no doubt that the war would be shortened materially. It was a prize well worth trying for; a bold, imaginative and adventurous gamble, and it so very nearly succeeded.

The operation was in two parts: a simultaneous airborne assault on the crossing places over the water obstacles on Second Army's axis – this was Operation Market; and a ground link-up by the divisions of XXX Corps, Operation Garden.

The airborne attack went in shortly after noon – Operation Market was reported to be going according to plan. As the gloriously hot September day wore on the Irish Group watched the huge air armada pass majestically and without opposition to the north.

Zero-hour on the Escaut bridge was like the start of some very improbable race, with the tanks head-to-tail, revving up their engines and waiting for the word to go. Promptly at 2.25 PM on September 17th Lieutenant Heathcote's tank advanced down the main road leading to Germany. He was keeping as close as he could to the barrage which was supporting the advance, a barrage from no fewer than ten field regiments, three medium regiments and one heavy regiment. In addition, overhead waited a 'cab-rank' of eleven squadrons of Typhoons ready to bear down upon any opposition.

All went well at first, and then before the astonished eyes of the very distinguished group of officers gathered to see the 'off', nine tanks of the leading squadrons disintegrated one after another in bursts of flame and shuddering explosions as the German 88s got to work. Even before the remaining tanks on which they were mounted could stop, the infantry of the 3rd Battalion were off their 'mounts' and into the ditches beside the road.

Now it was the turn of the Typhoons; over two hundred sorties went in that afternoon. The noise was indescribable: a barrage of 240 guns and more, Typhoons roaring in at zero feet with rockets blazing from all directions, tanks raking the surrounding countryside with Brownings and their main armament. Standing unmoved by the racket around him and the shots which were whistling about his ears, Colonel J. O. E. was clearly enjoying the excitement.

But the 3rd Battalion were not. They were itching to get at the Germans and out of the ditches. There was no mistaking the urgency of the moment, and when given the word to advance, they were off, one company on either side of the road sweeping outwards.

'I have never seen Guardsmen so angry, nor officers. The Krauts got rough treatment that day,' said an officer. The German paratroops couldn't take it; over 250 were captured and no one knows how many delayed their surrendering too long. All were sent back 'in very quick time'. Divisional Headquarters reported one lot as 'moving well' as they passed them six miles back – their escort was on a motor-cycle.

The Micks' blood was up; that, and the combination of close air support with intense and highly accurate artillery fire, cleared the opposition in record time. The advance continued. The bridge at Valkenswaard was reported clear and the group rumbled into the ruins of the town just as night was falling. The inhabitants turned out in force to welcome them. They were rung up by the German commander at Eindhoven, a further ten miles on, and told to 'hold on' – the telephone operator assured him in his best German that they would try to. But there were still Germans about Valkenswaard as well and a great deal of equipment. Four 88-mm guns were captured intact, and in endeavouring to destroy these an over-zealous officer managed to loose a round off closely over the heads of Group Headquarters – astonishingly causing no casualties. Those who were privileged to hear the subsequent explosion from the Irish Group Commander classify it as high on their list of treasured memories of the campaign.

They were off again early the next morning, this time with a squadron of Household Cavalry in the van. Again all went well initially, but then the whole group was held up beyond the little village of Aalst, still four miles from

Eindhoven. Rockets would have swiftly dispersed the defenders but 'no Typhoons available yet today', was the answer to this request.

Now the principal weakness of Operation Garden was becoming clear to everyone, for such was the country around this part of Holland that, abounding with streams and ditches as it was, the only route forward was along the main road – this was more often than not an embankment from which it was impossible for armour to get off. Second Army was in fact advancing on a one-tank front, and there was nothing they could do about it except smother opposition whenever it was encountered and press on, conscious that the leading tank was sure to go next.

Then suddenly German resistance collapsed and they could make full speed for Eindhoven. Waving at the cheering Dutch and Americans they passed on to Zon where Colonel J. O. E. crossed over the canal by boat to see his American opposite number. Apart from one company, who managed to set fire to a pig-sty and its inhabitants – by accident, they swore – the Irish Group spent a peaceful night on the south bank of the Wilhelmina Canal.

Working all night, the sappers managed to build a bridge in record time. By six o'clock the Grenadier Group who had passed through were able to rumble over the brand-new bridge and continue on their way.

The bridges over both the Maas and Grave and the Maas–Waal Canal were reported in our hands, and the Americans were said to be fighting only 300 yards from the main Nijmegen bridge. Arnhem was less happy; 'position obscure' was all they could get out of Intelligence.

The Grenadier Group closed with the US Airborne at Nijmegen where they were still fighting against a really stubborn enemy on the southern side of the bridge, and not much nearer than they had been in the morning. But even with this support they couldn't clear the way, so it was decided to send the 504th Regimental Combat Team of US Airborne across the river. Every available tank was lined up to give support and with the greatest gallantry the

Americans paddled their way across and fought their way up the far bank. Of the twenty-six boats used by the first wave only ten came back.

Nijmegen bridge fell later that afternoon, after the Grenadier Group had endured some of the hardest fighting of the campaign. Night fell but there was no rest for there were so many prisoners that escorts and protection parties occupied everyone fully.

The news from Arnhem was not good. The remnants of the British 1st Airborne Division sounded as though they were in real difficulties and must be relieved – quickly. The Grenadier Group had suffered heavy losses at Nijmegen and were in no fit state to take up the running again. The Irish Group were the only force who could do so, for it would be impossible to get up any other formation or group along the congested centre-line. At 11.30 AM on September 21st they advanced again.

It was the same story. Caught on a high embankment the leading tanks were all knocked out. The country was completely flat, offering no cover for infantry. The Typhoons were not forthcoming and all but one field regiment had been outranged by the swiftness of the advance. The attack ground to a halt; the great dash had failed only eight miles short of Arnhem.

43rd Division was passed through that night and managed to get two columns to the Lower Rhine thirty-six hours later, but elsewhere the Germans, recovering swiftly, mounted heavy attacks at various points along the attenuated line of communications of XXX Corps. The latter were so busy defending themselves that it was now no longer possible to continue the thrust.

The official report on the last Irish Group battle had said that the tanks had lacked vital support at a critical moment of the battle; but the truth of the matter was that the German resistance had been badly underestimated and the difficulties of the route for armour never truly appreciated. A Guardsman summed it up, 'We all felt pretty terrible about the unfortunate Airborne. Everything combined to stop us reaching them, but we tried so hard and got so near.'

But there was no rest. The Irish Group were sent to help look after the east flank of the Nijmegen bridgehead. The day after they arrived the Germans counter-attacked, using a mixture of units from both 9th and 10th SS Divisions – an enemy the Micks fervently believed they had seen the last of in Normandy. Again and again the close support of the tanks, and more particularly of the 153rd Field Regiment (the Leicestershire Yeomanry), by their incredible intensity of fire broke up the enemy attacks.

The Germans came back for more punishment the following day, and got it. They never tried again. The next day the 3rd Battalion were pulled out of the line and made their way to the rear. Many pets accompanied them. 'It would be cruel,' said Major J. S. O. Haslewood, as his Company Quartermaster-Sergeant nursed a wounded porker, 'to leave the poor thing to suffer in a lonely sty. Bring it along on a stretcher till we find it a home.' One of the 2nd Battalion scout-car drivers also took pity on a piglet, propped it up on the seat of his vehicle with its head and trotters sticking out through the roof, and drove back. A military policeman – an ex-Coldstreamer – noticed this peculiar pair on a bridge and asked what the passenger was supposed to be. 'A young Grenadier,' was the immediate response, and he was allowed to proceed.

The winter was a period of consolidation. Not only was it deemed necessary to clear the port of Antwerp as a forward supply base and relieve the greatly extended line of communications which stretched back through Holland and Belgium to the Channel ports, but also winter conditions made armoured operations practically impossible. Briefly the Irish Group came to an end and the two battalions led separate existences for a while.

Without warning on November 14th Colonel J. O. E. Vandeleur received orders to take over command on the next day of a brigade of 43rd Division. It was a terrible blow. Colonel J. O. E. had been an inspired and inspiring leader. His method of training, which was both imaginative and practical, had given the 3rd Battalion the perfect grounding for the Normandy campaign. Colonel J. O. E.'s own parting recollections were 'of the magnificent fighting qualities of the Irish Guards, and of the debt I owe them for their unswerving loyalty and gallantry'. 'In hunting parlance,' he wrote, 'it was like riding a patent safety, guaranteed to be in at the kill, however hard the going. In fighting it is the heart of the men that counts as much as the equipment and planning. That quality "heart" was never lacking.'

Major B. O. P. Eugster, MC, took over temporary command of the battalion until the arrival of Lieutenant-Colonel D. H. FitzGerald in February. Command of the Irish Group devolved on Lieutenant-Colonel G. A. M. Vandeleur.

December saw the Germans' final fling in the Ardennes, and this was succeeded by home leave, Brussels leave, conferences and field training for the two battalions. The weather improved, the news was good, and everyone enjoyed the first real rest, recuperation and leisurely refit they had had since D-Day.

Operation Veritable dawned on February 5th, 1945. Twenty-first Army Group's first spring offensive, it was designed to clear the Germans from the western bank of the Rhine, as forerunner for the great Rhine crossing.

But the planners had failed to account for the weather. The fields were sodden in the spring thaw, the roads proved quite incapable in these conditions of taking the heavy traffic involved and, to cap it all, the Germans released a great quantity of water from the Roer Valley; on the first day of the offensive the water rose eighteen inches, and continued rising.

Hopeless for armour, this was clearly an infantry battle, and 32nd Guards Brigade were ordered to reform as an infantry brigade. One disgruntled officer wrote, 'So while the 2nd Battalion lapped up oysters and did indifferent foot drill in Tilburg, we drove to a dirty barracks in Nijmegen.'

Hommersum, 'a beastly place it looked and was', turned out to be their objective. For two days the battalion sat around Hommersum, looking out across sodden fields. The slit-trenches filled with water, but an incessant German bombardment kept the battalion below ground and water level. The Coldstream and Welsh Guards managed to clear most of the surrounding countryside and 32nd Guards Brigade were just congratulating themselves that they were soon due for relief when the Irish Guards were ordered to clear a village some 2,000 yards ahead of the then existing line. They were informed that 'any Germans that were there have been washed away by the River Niers when it burst its banks'. The only possible lines of advance were two parallel tracks that did duty for roads in this part of the country.

The advance had hardly started before three carriers were blown up on mines and those that weren't soon became completely bogged. The platoons, more often than not knee-deep in water, came under heavy fire and it transpired later that there were no less than three German battalions holding the place. To add to their troubles the wireless sets were destroyed so they couldn't call down supporting fire. The companies, however, pushed on and rushed the German posts. 'The closer you got to their infantry,' said one man, 'the safer it was, as their gunners didn't try to follow.' They chased the Germans from ditch to ditch and house to house. When the leading platoon of Number 3 Company reached the village château they found it held by a battalion. Major D. M. Kennedy, MC, gathered up the platoon and swept it through the defences. 'How he did it,' wrote Captain C. D. Kennard, DSO, afterwards, 'beats me. They came under every sort of fire, from bazookas upwards. As he reached the château at the head of the platoon, Germans streamed out of the back door and hopped into trenches. "Come on, lads," he shouted, "we've got them now," and running to the trenches in front of everyone else he walked up and down killing Krauts with his revolver. He soon used up his ammunition, and was about to jump into the trench when a single shot got him from the flank. He was killed at once.'

On the left Number 1 Company were in difficulties in another village also held by a battalion of Germans. They, too, were caught in the open with no supporting fire and the only thing they could do was to pull back.

Number 3 Company pulled back in conformity. While Captain Kennard, the one remaining officer, concentrated the little parties of Guardsmen fighting their individual battles, Company Sergeant-Major Black supervised the evacuation of casualties. Although injured himself he personally rescued twelve wounded men from the open field.

The battalion had lost 175 men from Numbers 1 and 3 Companies, including both company commanders and all the platoon commanders. They pulled back weary and battle-sore to Nijmegen where they found the 1st Coldstream and their 2nd Battalion had the billets ready, 'with hot baths round the corner and dinners ready cooked'. They slept the clock round.

The Irish Guards were reformed at the end of February, by which time the advance towards Wesel was well under way.

At 5 AM on March 30th, 1945, the Irish Group drove across the Rhine and on into Germany, and almost immediately back into Holland. There was some fighting around Enschede and then they crossed the border again. In the customs shed the telephone was ringing. '*Wer ist da?*' asked a voice. 'This is your old chum Mick,' came the reply.

The road at Poppe was badly cratered, so they took particular delight in bulldozing the nearest house to fill in the hole. But in Gildehaus came the first serious opposition – from 8th Parachute Division, as fanatical as any Germans they had come across. The leading troop, driving down the narrow street past houses draped with white sheets and table-cloths, was ambushed; a platoon ordered to extricate them also suffered heavy casualties. The group withdrew to consider, and then went to sort the place out. 'We were in a merciless frame of mind when we arrived at Gildehaus,' wrote an officer afterwards. 'Typhoons raked a hill to the right and then Number 1 Squadron went in. The leading tank was knocked out at the far end of the town by an 88 mm. Its troop brewed up every house they could see.' Then it was the turn of the platoons and Numbers 2 and 3 Companies were ordered in. The Guardsmen were very angry indeed and 'laid about them vigorously by the light of the burning houses'.

On April 7th they crossed the Dortmund–Ems Canal, but there was opposition, of a sort, all the way. Furstenau, twenty miles on, saw the death of another company commander, Major M. Dudley. The 3rd Battalion then re-formed into three companies. Essen was reached on the 12th, and having outflanked the town they were about to set about clearing it when 5th Guards Brigade was ordered to switch to join XII Corps south of Hamburg.

The opposition seemed to be collapsing, but there were reports of one or two small groups still fighting stubbornly.

On April 19th reports came in that the Grossdeutschland Regiment had arrived at Zeven, a small town north of the Bremen–Hamburg autobahn. While a major attack was being mounted the Irish Group were ordered to send a company/squadron group to keep them quiet. The Official History takes up the story: 'Number 1 Squadron and Number 3 Company moved into Elsdorf and sent a troop and a platoon to occupy Wistedt, two kilometres to the west. They passed a pleasant evening shooting at transport flushed from Rotenburg by the 32nd Brigade, but the night was disturbed by the sound of troop movements closing in on them. The 15th Panzer Grenadier Division were retiring from Bremen. The morning of April 21st began with a heavy rain. At first light the troop of tanks in Wistedt moved out of the centre of the little village to cover the roads leading into it. Daylight came, the sections stood down and began to think seriously of breakfast. Out of the wood rolled two self-propelled guns; their first shots hit the tank posted as sentry on the road. Behind the self-propelled guns came a company of infantry. The tank went on fire as soon as it was hit and the crew baled out. Guardsman Charlton, the driver, stopped to look at the German infantry running down the road. He climbed on to the burning tank, unhooked the Browning machine-gun from the turret and jumped back into the road to meet the Germans. He faced them four-square, firing steadily. A bullet struck his left arm; he moved to a gate in the hedge and, supporting his arm on the top bar, still kept firing. His left arm was hit again, and he propped the Browning on the gate, firing and loading it with one hand. A final burst of fire shattered his right arm, and Charlton collapsed, the Browning on top of him. The Germans swept over him, but Charlton had ruined for them the effect of their sudden attack; the platoon and the other tanks recovered themselves. The Germans carried Charlton away, but he was already dying, and there was nothing they could do for him except bury him with the honour he deserved. A German officer who took part in the attack was later sent from a prison camp to the 2nd Battalion to show them Charlton's grave, as he talked so much about the bravery of an Irish Guardsman.'

This was the last major action fought by the Irish Guards in North-West Europe. On May 2nd the German Army in Italy capitulated; on May 4th the unconditional surrender of all German Forces was announced – it was to take effect from 0800 hours the following morning. But 0800 hours passed without anyone really noticing it. The war was over and had been for some time.

But there was one last occasion for the Guards Armoured Division, and on June 9th, 1945, at Rotenburg Airfield in Germany the division paraded for the last time with their armour. The tanks were drawn up with two battalions on the north side of the ground and two on the south. The saluting base was to the east, flanked by the infantry battalions and the men of the Services, while on the west side were the vehicles of the Supporting Arms. In the centre were the combined bands of the Scots and Welsh Guards.

At 11.30 AM Field-Marshal Montgomery arrived. After the general salute he inspected the parade, travelling in a half-track.

The party took their stand on the saluting base, and then the columns of tanks advanced and counter-marched across the ground, each tank commander traversing his turret and saluting. The columns turned and swung away to drive over the hills in the background, and as they disappeared the bands broke into the strains of *Auld Lang Syne*.

There followed a short pause. Then, coming over the crest of the hill could be seen the marching columns of the former armoured battalions. To their regimental marches, played in succession by the bands, they came striding down into the arena to join their comrades of the Foot Guards below.

As the old familiar tunes rang out on this, the greatest Household Brigade parade for six years, what memories must have glowed in the minds of the Micks there! Of Cagny; of the costly fighting in the bocage; of the wild scenes of liberation in Brussels; of the Escaut Canal and Hommersum. Of individual acts: of Lieutenant Lampard's dash at J. O. E.'s Bridge; of John Kennedy's last battle; of Guardsman Charlton's incredible personal battle for which he had just been posthumously awarded the Victoria Cross; and of so many more whose endurance and example and skill at arms had written one more passage in the fighting record of the Regiment.

So the parade drew to a close. The Guards Armoured Division was no more. For the first time for twenty-six years, a Guards Division was called to attention for the National Anthem.

In the Second World War the Irish Guards were awarded two Victoria Crosses, seventeen DSOs and eighteen DCMs. They had lost in all 734 dead, including fifty-nine officers; and their total of wounded was 1,578. About a third of the casualties of the Regiment in the Great War.

Chapter 14

Palestine 1947/8

The Irish Guards were permitted to keep their two battalions until, as part of the post-war rundown of the Regular Army, the 2nd Battalion were placed in suspended animation in June 1947. But by that time the 2nd Battalion were only at cadre strength and the 1st Battalion were on active service again – this time in Palestine under command of Lieutenant-Colonel D. H. FitzGerald, DSO.

The Palestine problem which the Irish Guards had first experienced in 1937 was brought to a temporary halt by the Second World War, during which both Jewish and Arab volunteers fought bravely and in large numbers for the Allied cause. But as soon as the war was over the whole question of Palestine and the Jewish State came to the forefront of world affairs.

The 1st Battalion Irish Guards reached Port Said in early March 1947. After a couple of days in the Canal Zone, they moved straight to Palestine in the most primitive train most had ever seen, with wooden seats and no doors, to Athlet some twelve miles south of Haifa. Here they joined the 6th Airborne Division, commanded by Major-General E. L. Bols, CB, DSO, Palestine veterans of nearly two years' standing and who had only recently left the intensely troubled area of southern Palestine for the slightly more relaxed atmosphere of the north, and Haifa in particular.

Duties in the Haifa area were certainly no sinecure, and as soon as they arrived the battalion were intimately involved in cordons and searches, curfews, vehicle escorts and road-blocks, and were in that state of perpetual vigilance which is the essence of Internal Security operations.

Haifa was out of bounds for much of the time, and the only occasion they actually operated in the town was when a curfew was imposed on the day of Dov Gruner's execution.* Otherwise routine duties and static guards were heavy, but no members of the battalion came under fire during their stay.

Their next destination was Khassa, twelve miles north of Gaza, and this was to be their home for most of the rest of the year. The camp was tented apart from a few semi-permanent buildings; stuck in the middle of a featureless plain, it was a dull place: 'like being in the bogs without the snipe', they reckoned.

But in late July, and to the delight of everyone, news arrived that they were to take up an operational role in Jerusalem, which was one of the hottest of trouble spots in Palestine.

The battalion were caught in the aftermath of the *Exodus* affair. For when that unhappy ship and her wretched cargo of immigrants were escorted back to Germany, the IZL and Stern Gang swore revenge. Isolated attacks were common. One Guardsman was killed in an ambush on the road to Khassa, and two others wounded in the same attack. Thus the Micks went to work with a will in their main task of searching for arms caches in the Jewish quarter.

The search for arms is an art in itself, involving patience, ingenuity, applied psychology and, in Jerusalem, a strong nose – for so pungent was the place that even the goats, they declared, had to come out of their owners' houses for a breath of fresh air. As on all such operations the experts soon emerged and some achieved quite a reputation.

When the Irish Guards returned to Khassa in early October they could look back on a reasonably successful time, for they had found five major arms caches and in-

* Dov Gruner – executed for his part in an attack on a police post. His death led to a wave of outrages by the Irgun Zvai Leumi (IZL), the extreme wing of the Haganah, or Jewish National Army.

numerable lesser pieces of evidence, as well as several 'wanted' thugs, a wealth of secret Haganah files and other important documents. They were sorry to leave.

After the New Year there were strong rumours that they were to move again soon. The barber's shop – always the source of the most reliable information – said it was to be Rosh Pinna in north-east Palestine, and on January 15th the rumours were proved correct.

The north-east corner of Palestine is one of the most beautiful parts of the country and particularly so in early spring, when the whole country is carpeted with flowers. A greater contrast to the drab flatness of the Khassa Plain could hardly be imagined. But there was contrast of another sort, too, for this was one of the most active sectors in the country.

On November 30th the General Assembly of the United Nations approved the partition of Palestine, and the entire Jewish community indulged in two days' riotous celebration. This was the signal for the outbreak of the inter-communal strife which had been simmering for so long, and in the first week of December, Jerusalem, Jaffa and Haifa witnessed numerous incidents of arson, looting and spasmodic exchange of fire. From then on it was the unpleasant duty of the British garrison to try to separate the contestants.

The principal road in the district ran north–south from Metulla, on the Syrian border, to Tiberias. Twisting through rough, hilly country it was wide open to ambush parties from over the Syrian border, four miles distant, or from the marauding Arab irregulars who eventually numbered close on 4,000 men and who trained and lived in the hills.

The battalion, again with 6th Airborne Division, now commanded by Major-General H. C. Stockwell, CB, CBE, DSO, an old friend from the Norwegian campaign, were in two camps built near Rosh Pinna on the Metulla–Tiberias road. To cope with their tasks, they maintained a 'Lightning' company, at thirty minutes' notice to move, and within that company a 'Lightning' platoon armed with a mortar mounted on the back of a fifteen-hundredweight truck. In addition a standing patrol of platoon strength in carriers with another mounted mortar was stationed in a wadi midway along the Metulla road. With this force they maintained some sort of peace from January 15th to the end of April 1948. It was a tricky task, but from the soldiering point of view a highly exhilarating one. For the battalion, with a total strength now of only four hundred men, it was also a very exhausting one.

Their own camp security was of special importance as a number of Army camps around Palestine had been raided (one cavalry regiment had experienced the embarrassment of losing an armoured car in this way). The camps at Rosh Pinna were never attacked, but alarms were frequent, and on one such occasion with the whole battalion standing-to, and rather looking forward to a proper battle, the Adjutant emerged from his tent to hear the Regimental Sergeant-Major giving final instructions to the perimeter guard: 'Now remember. If you see anyone up a tree shoot him. For no British soldier has a right to be up a tree after half ten at night!'

From then on incidents of varying gravity took place almost daily. A typical event occurred in early February and a platoon, sent to investigate an incident, were fired on by a party of Arabs who were beaten off after a sharp battle; it was in this engagement that Major P. A. McCall won a Military Cross for driving back an Arab force over a hundred strong in a spirited attack with fixed bayonets. A day or two later another platoon was cut off by a force of Arabs, this time some 300 strong, and only by some fast talking did the platoon commander extricate his men from that predicament. In this battle one Guardsman was killed and another wounded.

Also in Palestine at that time were the Regimental Band under their Director of Music, Major G. H. Willcocks, MBE. They had a tragic start to their tour, for shortly after their arrival the lorries in which they were travelling were ambushed near Jerusalem and one man was killed and five wounded – they were the first casualties ever experienced by the band in war. The band remained in Palestine until after St Patrick's Day.

Another sad event for the Micks occurred early in April, when a water truck was ambushed near Safad and the escort shot dead. The man who was killed was Guardsman Byrne. Byrne was a typical old soldier, the type who never gets the credit he deserves outside his own small community. It is the Byrnes of a battalion who are the sheet anchors, whose kit is always immaculate – for the simple reason that over the years they have acquired a complete duplicate set – and whose turnout is always beyond reproach. Byrne had been a duty-Guardsman for seventeen years and wished to remain that way. 'Big' Byrne he was always called; he was a huge man, six foot seven inches tall and broad with it, a fine heavy-weight boxer with fists like hams. He was a soft-spoken County Wexford man, and his death was taken hard by the battalion.

As the date for the cessation of the Mandate approached the tempo of incidents increased.

It was clear that neither side was prepared to listen to reason. Colonel Gordon-Watson thereupon announced that all British troops, with the police and Government officers, would be withdrawn 'today, tomorrow, the next day or next month' from Safad.

The Jews had tried to bribe Captain D. A. Lambert with £2,000 sterling for information about the exact date of the withdrawal from Safad; and they also received a rude answer from another officer to whom they offered £3,000 a year if he stayed behind to teach them street-fighting! But on the afternoon of April 16th Safad was evacuated.

During the earlier part of the day all spare men and equipment had been surreptitiously removed. Then, on a given signal, every man who up to that moment had been looking completely unconcerned jumped into his previously allotted place in a vehicle and they were off. All clear of the town in under five minutes, the last they saw of Safad was a mad dash being made by the Arabs and the Jews towards the recently evacuated police station which dominated the town.

The Mandate finished on May 15th and civil war raged in its stead, but this, happily, was no concern of the British troops still in Palestine. The battalion were in camp near Haifa except for one company situated somewhere in the clouds on Mount Carmel. Here they lived a delightfully detached life until they returned to ground level to be greeted by the Sergeant-Major at the barrack gates with, 'Well, you little angels, I will see you on the square tomorrow morning.'

At last the day of departure arrived, and on June 15th, 1948, the Irish Guards embarked on HMT *Empire Test* bound for Tripoli. There had been heart-searching to the end, for their faithful mascot Capstar the donkey, and his equally cherished companion, a goat, had to be left behind.

After the Palestine campaign, the battalion were awarded one OBE, to Colonel Gordon-Watson, a Military Cross to Major McCall and two British Empire Medals.

Chapter 15

The years of change

The battalion remained in Tripoli for a little over a year, until in April 1949 they set sail for England and Public Duties. In the previous year the Household Brigade had returned to ceremonial uniforms, so it was to an endless round of tunic and bearskin cap fittings that they came back, though only the oldest members of the battalion had ever worn them before.

The year 1950 was the Golden Jubilee of the Irish Guards, and Micks from far and wide gathered to celebrate it, including the Colonel of the Regiment, Field-Marshal Earl Alexander, then Governor-General of Canada. He had succeeded Field-Marshal the Earl of Cavan to the Colonelcy on the latter's death in August 1946.

The early 1950s saw the 1st Battalion Irish Guards undertaking tours of duty in Germany and the Canal Zone of Egypt, when a number of officers were seconded for duty in Kenya at the later stages of the Mau-Mau emergency. The Household Brigade were not called upon to fight in Korea, but contributed a brigade in the Malayan emergency. During the 1956 Suez crisis only the Guards Independent Parachute Company were involved – whose forerunners had been 1st (Guards) Parachute Battalion, dissolved after the Palestine campaign – and they were among the first to land on that expedition; this party included a number of Irish Guardsmen. It was ten years to the week from the end of the Palestine campaign of 1948 before the 1st Battalion Irish Guards were to embark on active service again.

The 1958 sports season had already started and the athletics and shooting teams were away training hard at Pirbright when, on June 13th, and without any warning, the Commanding Officer, Lieutenant-Colonel J. W. Berridge, MBE, received a signal placing the battalion at seven days' notice to move, to an unknown destination.

From then on activity was feverish; the two teams were hurriedly called back to the battalion and incorporated once again in their companies and platoons; the Orderly Room started the seemingly endless task of form-filling, identity discs were stamped and issued, parcels of civilian clothes wrapped and addressed, ammunition issued, checked and sorted: all the preparations for a sudden change from peace to war. There was still no official news where they were going, but a large consignment of 'Hats, Tropical' which appeared one day in the Quartermaster's stores could only mean one thing – the Middle East where yet another crisis seemed pending, or perhaps Cyprus where EOKA – the military arm of Enosis (Union with Greece) – was still very active.

Cyprus it proved to be, but only as a springboard should trouble brew elsewhere in the eastern Mediterranean. The Parachute Brigade, the first-line strategic reserve in England, were already committed and had flown off; Headquarters 1st Guards Battalion and the 2nd Battalion Grenadier Guards shortly followed. The Micks flew on June 20th. There were a good number of men due for discharge before the end of the year, and who could not be taken. With their loss the battalion was well below strength. So to make up the establishment a company of the Welsh Guards joined the battalion under the command of Major P. P. De Zulueta. That company, known as Number 3, remained with the battalion throughout their service in Cyprus; it was a thoroughly happy union.

The battalion had all arrived by June 24th. They found themselves in Ascot Camp, an area just outside Nicosia which in July resembled desert, with a single tap – and that didn't work – to provide the needs of 800 men. Their stay was mercifully short, and on June 29th the companies moved out to be employed on Internal Security duties in the rural areas around Nicosia, an operational area some 500 square miles in extent.

Their duty was to instil confidence in the civilian population and in particular to prevent inter-communal disturbances. To achieve this each company was given its own sector to look after, and, in close cooperation with the armoured cars of the Royal Horse Guards (the Blues), attempted to control it. A show of military force was sometimes salutary, but merely the presence of troops actively patrolling was enough to keep the temperature down. Each company had a few pipers with them and their performances were invariably received with wild enthusiasm.

Cyprus in mid-1958 was still simmering, but the clashes of the previous two years had not been repeated. It was clear that a political settlement was imminent, but even so EOKA were far from subdued. For the most part EOKA activities were confined to clandestine operations – moving in arms and the like – but they were a viable terrorist organization and any lapse on the part of the Security Forces would undoubtedly be punished. It was a game of trying to keep one step ahead of the enemy all the time and of a continual grind of road-blocks, vehicle searches, cordons and curfews. From time to time some information came in and was immediately acted upon, with mixed success.

It was exhilarating in the rural areas around Nicosia. But in the Middle East the situation was no nearer solution and it seemed as if the Parachute Brigade, which had been suddenly flown to Jordan, were in for trouble. So the months of July, August and September saw the Irish Guards standing by at varying degrees of notice to fly to any part of the Middle East.

The uneasy truce on the island broke in the first week of October, and EOKA started up another series of savage attacks; the Director of Operations himself narrowly missed being blown up and a number of ambushes took place. Then, on October 3rd, two British women were brutally shot down from behind in Famagusta. The atmosphere was tense; a major outburst of violence was expected hourly and the battalion stood by to move into Nicosia itself. But the excitement died down and the round of sports and training returned.

On October 19th they took part in the cordoning of a village in the Kyrenia foothills; a task which kept them occupied for a week. A number of terrorists were identified by the hooded informers who sat in an armoured car and peered through the gun ports nodding or shaking their heads as each villager passed before them. The only casualties in the battalion were those who ate too many unripe figs.

At the end of this operation they went back to Nicosia and waited. Mysterious signs called 'Queen Victoria Memorial' or 'St James's Palace' gave credence to the rumour that they were soon going to England and Public Duties again. And in early November, rumour became fact. It was just in time, for as they left the weather broke. The old canvas, a relic of Canal Zone days, was rotten and leaked; the camp area was soon a muddy mess. But this was now no concern of theirs.

They went to Windsor this time, and were to remain in England for the next three years. These were formative years in the evolution of the Household Brigade, for in January 1960 the Junior Guardsmen's Company was started; this was the modern counterpart of the band-boy system of another age. Under the new scheme boys of 15½ to 16½ could join the Brigade and be trained first as soldiers and then in a trade of their own choosing.

The year 1960 also saw another landmark, for it was then that military economics forced the merger of the Guards Depot at Caterham and the Guards Training Battalion at Pirbright; the new Guards Depot was to move to Pirbright. The closing of Caterham ended an association with the Household Brigade which covered eighty-three years. To many old Guardsmen Caterham *was* the Guards;

and that dirty yellow building seemed to epitomize to many all that the Brigade meant to them and their forebears. The reasons for the closure were sound enough and brought about by the gradual shrinking of our forces in line with the reduction of our overseas commitments; nevertheless it was a wrench, for Caterham and the Caterham tradition had seemed sacrosanct.

In November 1961 the battalion moved to Germany to take up duties with 4th Guards Brigade. This was their first introduction to the APC – the Armoured Personnel Carrier – with which rifle companies were equipped. The new vehicle gave them a mobility hitherto undreamed of, but it laid great strain on the maintenance, driving and signalling capacities of the battalion and entailed endless courses to train enough of these specialists.

The period from 1960 to the departure of the 1st Battalion Irish Guards for Aden was perhaps the most successful time in sport that the Irish Guards have ever achieved. In 1960 they attained the long-elusive goal of winning the All-Army Boxing Championships, and then proceeded to win it on three more occasions in the next six years, being runners-up once and semi-finalists in the other two years. These were results to gladden the heart of any Mick, but each round of the competition was a nerve-racking performance, for Guardsmen rarely come in the lighter weights. Time and time again they lost the first three or four fights of a scheduled eleven-bout event, and on two occasions were dormy five and in the seemingly impossible position of having to win all the remaining fights to survive to the next round – and an unlucky disqualification can happen to anyone.

In their other main sport, athletics, they also achieved their highest ambition, to regain the Army Championship which they had previously won in the year of its inception in 1920. Year after year they had reached the Army Finals and found themselves fourth, fifth or sixth of the eight teams competing. But in 1964 they were runners-up to the Cheshire Regiment, their great athletics rivals; and, in

1965, gained their revenge and carried off the trophy. Lesser successes were scored in other sports. The Micks, purists may complain, have a technique which is sometimes a little rough, but they make up for this by fighting spirit – and it is this spirit that pervades all they do.

In 1963 the battalion were asked to provide one full company for attachment to the 2nd Battalion Scots Guards, who were then in Kenya, for the Scots Guards were temporarily finding difficulty in getting enough recruits to keep their two battalions going. And again in 1964 a second company was sent to the 1st Battalion Scots Guards, then in Malaya and bound for Borneo where the Irish Guards were able to cement the recently announced alliance with the 4th Royal Australian Regiment.

Soon, everyone felt, it must be the turn of the Micks to go abroad, and they were delighted to a man when they heard they were to relieve the 1st Welsh Guards in Aden in October 1966. From that moment, training took on a fresh sense of purpose.

But 1966 was memorable in another way, for new Colours were to be presented that June and they were to be Trooped at the following Birthday Parade.

As soon as St Patrick's Day had passed, the Escort for the Colour and Number Two Guard was picked, and a large number of reserves earmarked. From then on, everything was subordinate to the preparations for the Presentation of Colours and the Birthday Parade itself. Companies were denuded of non-commissioned officers and Guardsmen of the right height and of sufficiently immaculate bearing, and most days saw the chosen teams parading under the eye of the Adjutant, the Regimental Sergeant-Major and his henchmen. April passed, and most of May – they were getting good, very good indeed – and the pessimists pronounced that it was too soon. For it is a delicate business. A racehorse trainer endeavouring to produce his charge at the peak of condition on a certain day has an easy task by comparison with that required to prepare a corporate, thinking collection of individuals to the heights of perfection and beyond in close-order drill. Taut nerves and staleness are the two contagious diseases which can reach epidemic proportions if not handled with exquisite care. There was always the comforting knowledge that Irishmen invariably rise to the occasion, a trait which exasperates those who do not understand them; nevertheless, as a Mick of another age said of another parade, ''Twas a breathless performance.'

There was an added complication, too, for the Presentation of Colours was to take place in the grounds of Buckingham Palace and on grass, a totally different medium to work on.

Gradually – imperceptibly, the pessimists said – the Escort for the Colour and Number Two Guard improved. The first rehearsals passed satisfactorily, with no major 'bogs' and with sufficient room for improvement to satisfy the pundits – who aver that no rehearsal must ever be perfect.

A couple of practices on the lawns of Buckingham Palace to condition everyone to the form of parade, and a dusty session in the Riding School as a contingency against wet weather, and Friday, June 10th, was upon them.

It was a perfect day for ceremonial – fractionally cloudy, with promise of sunshine ahead, and a good breeze to keep the bearskins ruffled and their wearers cool.

It would be difficult to imagine a more lovely setting for such a ceremony: the grey façade of the Palace, the broad sweep of the lawns, and across the lawn facing the Palace the battalion drawn up resplendent in their scarlet tunics and bearskin caps, with their plumes of St Patrick's blue. To their right was the Regimental Band, and behind them the Pipers with their saffron kilts, green tunics, and their long green cloaks hanging over their left shoulders, the true historic dress of Irish pipers.

There can have been few dry eyes at that most moving and poignant moment in all ceremonial drill when the Old Colours are marched off in slow-time to *Auld Lang Syne*.

Former Micks were present in full force; it was a truly family gathering, and an Irish family gathering is an occasion apart.

That afternoon the Queen joined her Irish Guards in a garden party in the grounds of Chelsea Hospital, and a memorable occasion became more memorable still.

The Birthday Parade the following day was an unqualified success. 'Best Escort I have ever seen' was the comment from the Old Comrades and friends of the Regiment – but then they may have been a little biased.

Within six weeks of that great day the battalion had finished their last spell of Public Duties. A fortnight's brisk training in Wales to condition them to the hills of Aden, some embarkation leave, and they were off. It was a drastic readjustment to move from the conformity of the greatest military ceremonial parade to that branch of modern soldiering which calls for individual initiative of the highest order – namely Internal Security.

Chapter 16

Aden

By the end of October 1966 the battalion had completed their fly-in to Aden from England. Seemingly inevitable delays had dogged this move – a familiar experience for military air travellers – but despite some vicissitudes the 1st Battalion Irish Guards, commanded by Lieutenant-Colonel J. A. Aylmer, gradually assembled in Aden – although in by no means the order they had started out.

The battalion, nearly eighty per cent Irish, were just over 600 men strong; little enough for such an exacting and wearying task as Internal Security. There was still a full company of the Regiment serving with the 1st Battalion Scots Guards in Malaysia so, as a gesture, and to make up the numbers, a platoon of Scots Guards were attached to the Micks. In addition, instead of their customary role of stretcher-bearers, the Pipes and Drums were used as a full fighting platoon. The most revolutionary change of all was the incorporation of an air platoon of three Sioux helicopters with pilots and supporting ground staff, and Captain G. M. Daintry had the privilege of being the first Irish Guards officer to fly in support of his own battalion on active operations.

It was a strange situation they found themselves in. 'The Emergency which never was', someone described it. For at that time, in late 1966, tempers were cool and the terrorists were still preparing and arming themselves for the holocaust which was to break the following spring.

In the back-streets of Big Aden and the outlying suburb of Sheik Othman patrolling continued, inviting grenade attack and relying on the superior weapon-handling of the Security Forces to come out on top. A 'tangible presence' was how it was described; others called it Operation Aunt Sally.

The most glaring omission was the almost total lack of routine intelligence. Early in 1966, by a series of brilliantly organized assassinations, the cream of the local Special Branch had been eliminated, and it had not been possible to replace that organization in anything like a viable form.

It is not the purpose of this book to decry the prevailing atmosphere in Aden in 1966 when the Irish Guards arrived there; suffice it to say that British troops can seldom, if indeed ever, have been placed in such a position of powerlessness, handicapped by the lack of sympathy from a civil authority which made nonsense of their supposed task, unsupported and misunderstood at home and placed in an invidious position from the start.

So it was not without relief that the battalion deserted the smells and dust of Aden for the different smells and greater dust of the hinterland, the Radfan, as it was known.

Habilayn lies sixty miles north of Aden. The road is tarmac most of the way, but over the last stage it winds through the Wadi Matlah, an imposing gorge ten miles long, flanked by craggy, volcanic rock, the whole a heartless wilderness and a heaven-sent ambush position. Working in the stifling atmosphere of the wadi was a small party of Royal Engineers and a larger number of local labourers striving to repair the ravages of a torrent which had swept down the gorge that spring, brushing aside the puny culverts so painstakingly built the previous year, and in the process hurling aside into irreparable ruin a sixteen-ton bulldozer that happened to lie in its path.

To guard the Sappers in their labours Number 3 Company were based on a camp at the foot of the wadi. Battalion Headquarters with Number 1 Company were at the main airbase of Habilayn, with a squadron of armoured cars and a company of the Parachute Regiment under command. Further north, a few miles from the Yemen border itself, the remaining rifle company, Number 2, luxuriated in the sometimes delightful climate of Dhala.

It wasn't peaceful, however, for at ten o'clock on the night of December 30th, and with no previous warning, a shattering explosion occurred near the perimeter wire of the Number 3 Company camp at the Wadi Matlah. This was immediately followed by heavy small-arms fire, blindicide (a form of rocket launcher which hurled rockets some nineteen inches in length and three inches in diameter a distance of two hundred yards or more – a lethal and effective weapon) and mortar fire. The fire was returned by the pickets, but to the anxious listeners on the Battalion Command wireless network it was clear that this was no haphazard attack. Well organized and well conducted, nothing like it had occurred in the Radfan since the days of the main campaign in 1964.

Captain D. B. W. Webb-Carter (son of Lieutenant-Colonel B. W. Webb-Carter who had been commanding the battalion of the Duke of Wellington's Regiment which had finally relieved the remnants of the 1st Battalion Irish Guards at Anzio in 1944), then temporarily in command of the company, was the mainspring of the defence. It is always a nervous moment when troops are under fire for the first time, but the concern was needless: the modern Mick behaved as had his forebears and with the same humour, the same spirit and the same apparent relish.

The attack lasted for over an hour and a half, a prodigious time by Radfan standards, until the dissidents withdrew and vanished whence they had come.

The sole casualty was the company refrigerator which, shot through the guts, gave a despairing noise of rending metal and worked no more. And the next day a West African driver who had been attached to the company gave notice, rightly feeling that this was no place for a dusky face at night with itchy-fingered Micks about. For his fine leadership during this attack Captain Webb-Carter, who had only arrived in the country a few days before, was awarded the Military Cross.

It was next the turn of the Dhala garrison. The beauty of the position aesthetically was more than counterbalanced by the vulnerability of their camp. Sited on a long narrow ridge, it stuck up like a ship on a plain as flat as any sea. On the ridge itself there was hardly any room left for another tent. An incendiary bullet would have cleared the ridge of canvas in no time; the fire risk was appalling.

The Dhala valley was immensely fertile. The climate was delightful, the rain frequent, the soil rich. The Emir of Dhala, residing in his palace of glaring white, extorted tolls from those passing through his dominion. His subjects and other travellers passed at will up to the very gates of the camp, all bearing their weapons, as was their wont. It was a simple matter for some of these supposed innocents to lie up till dark or to mine or booby-trap while passing.

Next door, some four hundred yards away, was a camp of the Aden Regular Federal Army commanded by a Colonel Yafai, who was in over-all command of the Dhala garrison, and with whom relations could not have been more cordial. As night fell each camp was placed in a state of instant readiness. They had two principal opponents, Ali Antar and Ali Actar, who attacked in turn with a regularity which was almost predictable. Fortunately the dislike for the Emir by these two was exceeded by their hatred of each other, so they never joined forces.

Nevertheless, the military received their share of attention and on New Year's night it was the turn of Number 2 Company. It was pitch black, with hardly a star in the sky, and under cover of the darkness the enemy must have crept close in upon them, for suddenly at ten o'clock a wicked fire was opened with blindicide and machine-guns. It was a great battle while it lasted, but with no casualties on either side, except for the company ration store which received a direct hit and was the cause of an unscheduled visit by the Quartermaster who came to verify the monster grocery order which was submitted after the affray. Sergeant Johnston was subsequently awarded the Queen's Commendation for Bravery for his gallantry in bringing a machine-gun into action under heavy fire.

The battalion's role up-country was partly to protect the installations there, but also to provide 'Hearts and Minds' patrols for the less accessible regions. For these forays the helicopter was essential, and liaison with the RAF and the Army Air Corps throughout the up-country tours was the most important feature of their activities. The helicopters were in constant use either for ferrying troops, or for supplying them in their forward positions, or for conveying commanders from one area to another. It was a new medium of war, but one they found absolutely essential in this inhospitable terrain; to command operations from the air or to establish aerial pickets to cover some approach march requires a speed of reaction far removed from the pedestrian thinking of normal infantry soldiering; it was a highly instructive experience for everyone.

The big operation which was to be the swan-song of the first tour up-country was appropriately named Tara Trot. Battalion Headquarters and two companies, with a troop of the Queen's Dragoon Guards in support, plus a twenty-five-pounder gun of the 1st Regiment Royal Horse Artillery, were involved; the venue, the notorious Wadi Taim, an area which had required a full brigade to suppress it in 1964 and which marched with the most impenetrable of dissident-held territory. Dragging the coat it was intended to be, and dragging the coat it was – the insult to rebel *amour propre* could only be expiated in one way. They obliged in full measure.

On the very first night the rebels attacked both the main position held by Number 1 Company, and another further up the wadi held by Number 3 – but with no success. This, however, was only a preview to sound the general alertness. The true test followed on the next night. Battalion Headquarters, lurking in the rear of Number 1 Company, received the brunt. It was a perfect night for attack. The moon in its last quarter, rising late and setting around midnight, provided just enough light for the rebel approach march and better still for their hurried scuttle to safety before dawn and avenging air-power could catch them.

The command post was just settling down to a nervy night when their peace was rudely disturbed by the blast of a blindicide rocket. It happened to be aimed at the latrine, occupied at the time by the Signals Officer. It is declared that he holds the record for the sitting-start dash. Battalion Headquarters then found themselves in the forefront of the battle.

Somehow the attached troop of armoured cars of the Queen's Dragoon Guards managed to make their way over the rocky ground and helped to beat off the attack. Blood was found in the rebel positions the following morning, but they had borne away their wounded. Less lucky was a rebel who was a member of a party attacking the other camp; as he was firing his blindicide in that attack, a chance shot went down the barrel of his rocket and blew the bomb and him to perdition. That attack broke in disorder and the survivors vanished leaving a trail of equipment behind them.

Such was life up-country. Patrols by day and by night, occasional company-size operations. Frequent attacks to add spice to the routine, no casualties to the defenders, a few for the attackers. Ten o'clock at night to one o'clock in the morning were the key hours. After that everyone could go to sleep confident that peace would reign until the following night.

It was a fit, hardened but tired battalion which returned to the different battles of urban Internal Security at the beginning of February, in time to witness and celebrate the victory of their boxing team in the Middle East Command Championships.

Trouble was brewing on the labour front in Aden. The first of many General Strikes was called for February 11th. Little Aden and the road to Aden itself were the battalion responsibility. There they had none of the trouble that occurred in Big Aden. In Sheik Othman the situation was more tense and Number 2 Company was sent to assist. They caught one grenade thrower. He had the temerity to throw his grenade at the armoured car in which were a section of the company. This was too much for the temper of one Guardsman, and with a turn of speed which astounded his friends he went in pursuit and managed to shoot down the fleeing Arab as he flitted round a corner. Peace, or comparative peace, returned in time for them to celebrate St Patrick's Day in the proper manner.

The Colonel of the Regiment, Field-Marshal Alexander, arrived the day before as eager as ever to see the kind of war they were fighting, particularly since before the Second World War he had commanded a brigade on the North-West Frontier of India, an area of very similar terrain. Under the eyes of the patrolling helicopters and pickets posted on the surrounding hills the notables assembled. The parade, despite a certain lack of practice, reached its usual standard. Once again fresh eyes had seen that transformation from battle drill to an immaculate parade-ground appearance which is a hallmark of the Brigade of Guards.

That very day it was announced that the Irish Guards were not to go into Big Aden as planned and that another up-country tour was their destiny. They were not sorry. It was Number 1 Company's turn to go to Dhala this time; they had rather more than their share of excitement, for on the day of their arrival the Commanding Officer of the Federal Regular Army garrison was blown up by a mine, and they were almost nightly under some sort of attack, one tent, luckily unoccupied, receiving a direct hit from a mortar bomb. It had been one of the unwritten rules that the anti-tank gun should not be fired at the dissidents; the reason had no tactical significance, it was just that on a previous occasion the noise of the big weapon had so surprised one of the many wives of the Emir of Dhala that she had suffered a miscarriage. But after the latest batch of attacks the rule was rapidly rescinded.

Shortly after the return up-country the Scots Guards

platoon, which had come out with the battalion, departed, to be replaced by one from the 1st Battalion Coldstream Guards. They arrived in time to be impotent witnesses to a tragic yet gallant affair.

Intelligence up-country was scanty, and of dubious reliability. There had been rumours of an impending attack on the main camp area at Habilayn for some time, and each night a number of patrols had been sent out to ambush likely approach routes. But on April 20th the rumours were a little stronger than usual. As darkness fell three platoon-strength patrols stealthily moved out of the camp and to their prearranged positions; careful artillery and mortar support was mapped out.

First one and then the second and then the third reported themselves in position. The lights at the main camp at Habilayn dimmed and went out. Activity was minimal except among the mortar and gun crews who were warned and waiting.

If the enemy were to attack the camp at midnight, which was the latest they could afford to and get away before dawn, they should pass the ambush positions by 10 PM and no later. But 10 PM came, and then 10.30 and still there was no news. They must have missed them, or else they weren't coming that night. Then, at 10.40, as everyone was beginning to relax, came the report that heavy firing had been heard to the north-east. An urgent call came from Number 5 Platoon's patrol for the prearranged mortar fire task, and then all was silence over the air.

This patrol, led by Lieutenant R. J. S. Bullock-Webster, with Sergeant Thunder as his platoon sergeant, knew the drill. Their move into position had been accomplished silently and effectively.

The moon kept appearing and disappearing in the fleeting clouds. There was no sound. It was a completely still night. The moonlight threw queer shadows against the rock and against the wadi path where they watched.

The patrol was in two parts, commanded respectively by the platoon commander and platoon sergeant; rendezvous points and alternative ones had been carefully worked out and the ambush layout had been planned and rehearsed many times.

Precisely at 10.35 someone heard a movement to the right; the message was passed on, each man touching his neighbour with his foot until the whole patrol was alert and watching the path. Some hundred yards away they could see a body of men in uniform moving their way. Fifty yards to go before they reached the chosen killing ground, then forty, then – at the very last moment the dissidents veered off.

It was a terrible moment. Where had they gone? Why? They weren't kept waiting for long. The rebels had indeed veered off course, up another path, it later transpired, but then they turned back again, and right on top of the waiting patrol.

It was close-quarter fighting at its fiercest. On first contact the rebels had loosed off their blindicide rockets, carried in their front group, with fearful effect: two of the three radios were put out of action immediately, the other was destroyed moments later, and two men were literally lifted off their feet by the blast and killed. The rebels bravely pressed home their attack.

Lance-Corporal Lewis was on the outer flank of the patrol position and had been forced to spring the ambush just as the enemy were on top of him. He then proceeded to stand up in full view of them and engage in a point-blank fight; this incredibly brave act was an inspiration to his whole section and they stood their ground despite being heavily outnumbered and outgunned. Guardsman Bell in particular engaged in a hand-to-hand combat with the rebel leader and was killed in the fray, but he killed his opponent too.

When the patrol commander finally ordered the position to be abandoned Lance-Corporal Lewis covered the withdrawal of his section, while the platoon sergeant did the same for his part of the patrol, and was killed holding off the enemy. The patrol were scattered into two parties by now and they made their separate ways back from their respective rendezvous points to the main camp. Here it was discovered that four men were missing. With great gallantry Major G. J. B. Edgecombe of the Army Air Corps and the Company Commander, Major S. J. B. Blewitt, flew over the area in a helicopter to try to spot survivors, and then landed to pick up the dead.

The patrol had killed four of the enemy, including the gang leader, a notorious rebel, and wounded seven others, but their own casualties were four killed and one slightly wounded. It was fitting that for his fine courage and leadership in this patrol action Lance-Corporal Lewis, who had been promoted Lance-Sergeant in the field the very next day, was awarded the Military Medal, and Guardsman Bell, a posthumous Mention in Dispatches.

It was a veteran battalion which returned to the sordid work in Aden once again. Since they had left, the situation there had been deteriorating steadily and was now in danger of getting out of hand. With the evacuation announced for August it was clear that they were in for a busy and unpleasant time.

The six-day Arab-Israeli War had started on June 5th. Everyone had fondly hoped the humiliating defeat of the Egyptians would quieten Arab fervour, but the reverse happened and Aden was in ferment. The battalion were ordered to undertake the protection of the refinery in Little Aden, which held most of the refined oil reserves in the Middle East. This task occupied them fully for some days. Meanwhile, they always had in the back of their minds the other eight contingency plans which might have to be put into effect should the situation in any of the sheikdoms down the coast deteriorate, or look like deteriorating sufficiently to justify intervention. On the day of the Queen's Birthday Parade in London, it happened.

Colonel Aylmer takes up the tale: *

'1st Battalion Irish Guards Detail for Saturday, June 10th, 1967, Queen's Birthday.

* Reprinted by permission of the *Household Division Magazine*.

'Saturday, June 10th, 1967, will be observed as Her Majesty's Official Birthday. The Internal Security situation will not allow any form of parade, but those not on duty in the battalion may observe Sunday Routine. A broadcast of the Queen's Birthday Parade in London may be heard on the British Forces Broadcasting Service at 1535 hours.'

Here is the Commanding Officer's diary of events for that day:

0100 Returned home after visiting sentries on BP and Shell Refineries in Little Aden. Some drama over some missing reliefs in one of the platoons, but happily solved.

0800 Lovely late breakfast on Sunday Routine. Ask my Orderly to get in champagne supply for a small party this evening to celebrate Queen's Birthday.

0845 Arrive Orderly Room. Adjutant passing out drill course (potential NCOs) on square. All serene.

0930 Dutifully reading minutes of last meeting of Regimental Lieutenant-Colonel's Conference about dress in London.

0935 Radio message from Number 1 Company Commander commanding our Internal Security Force. He tells of strike in Little Aden, BP workers walking out, trouble expected.

0945 Strengthen stand-by force for Internal Security and call out a Queen's Dragoon Guards troop to help.

1000 Peace again. Resume London reading.

1015 Call from our Brigade Major – State 'Red' – rioting in Aden. Our emergency Internal Security plan put into action and Number 2 Company called out to cover Little Aden–Aden road.

1020 Adjutant still passing out drill course on square; looking rather hot now.

1040 Another call from the Brigade Major. 'The Residency at Mukalla (400 miles down the coast) reported under heavy fire. Can you put Operation Holmes into action at once?' This is an operation to rescue residency staff from Mukalla for which we are at twenty-four hours' notice. (Having served in Middle East Command for six months one knows twenty-four hours' notice may often mean minus five minutes, so we are moderately prepared.)

1044 Adjutant disperses drill course much to their relief and astonishment. Order Number 3 Company, whose last platoon has come into camp off Internal Security Guards fifteen minutes before, and Tactical Battalion Headquarters to get to Aden Airfield with their kit as quickly as possible.

1045 Order Number 2 Company to turn round in mid-road and return to stand-by. Tell Brigade Headquarters they must relieve us of command of Little Aden Internal Security if they want us to get to Mukalla. Brigade Major sounds a bit hurt, but agrees helpfully and immediately.

1046 Heart-rending appeal over the radio from Number 1 Company Commander for more troops as riot forming up in Little Aden. Callously say 'nuts' but ask Brigade Major to provide the troops from elsewhere. He sounds a bit more hurt.

1100 Set about getting my own kit ready. Hampered by absence of Orderly in NAAFI buying champagne. Luckily he returns with the festive drink.

1115 Call at Brigade Headquarters. Am told there is also rioting in Sayun, about ninety miles away from Mukalla, and I may have to deal with both . . .

1140 Arrive Falaise Airfield (the local airstrip).

1155 Mount the tiny Beaver aircraft. The Brigade Major in London must be mounting his horse.

1450 Mukalla in sight at last.

Luckily the Residency was standing 'as four square and Victorian' as ever, but the situation at Sayun was dubious and reinforcements were urgently required.

1515 Land Riyan (an airfield some 15 miles from Mukalla).

1525 Joined there by the first Wessex helicopter loads of Number 3 Company.

1530 Decide to get most of Number 3 Company to Sayun before dark. Ghastly time/distance/fuel/passenger and load calculations follow. Eventually give orders to send the two Wessex with troops direct to Sayun and one Beverley with troops, kit and helicopter fuel to Ghuraf, the nearest airfield some fifteen miles away.

1603 Leading Beverley arrives – chaos sorting our bodies and kit all over the tarmac (we are still on peacetime accounting).

1643 Second Beverley arrives – more chaos – heat 104 degrees, humidity 76 per cent.

1645 Third Beverley arrives – even more chaos. One Company Sergeant-Major heard to remark, 'They are just coming up the Mall now, sir.'

1700 Wessex load with platoon ready for departure.

1705 Take off and fly for an hour over amazing bleak high flat and barren country split by sudden sheer canyons. None of the pilots have flown this route before so some concentration on navigation.

1805 Over Sayun. Realize I am not sure where Wessex should land. Wessex pilot reports a riot – we fly over it and it turns out to be a very English-looking football match, but unfortunately where he had hoped to land.

1810 See a barrack-like area with soldiers. Decide to try and tell Wessex to land. Soldiers look bewildered – awful moment – will they be hostile? We overfly, our gunner's finger itching on the trigger.

1811 Micks out of Wessexes in a flash and straight down into fire position. Arab officer approaches, platoon commander gives thumbs-up. Great relief, they are friendly.

The 'soldiers' were the Hadramaut Bedouin Legion, a fine and highly trained force, who had they been hostile would have been a very formidable enemy indeed.

The sudden presence of the troops in the area had the desired effect, and from that moment peace was restored. Their most nervous moment was when it was decided to reopen the British Bank in Mukalla, as the Resident expected an immediate run on the bank and ensuing riots. To prevent this a platoon were in the air in helicopters at the appointed opening time. And the Commanding Officer, determined not to miss this bit of fun, was standing by in another helicopter hovering over the dome-like top storey of the bank building with a strong man and a stout rope ready to haul the bank manager to safety should he be chased to the roof by his furious customers. Yet another use for the versatile helicopters.

The situation had calmed sufficiently for Battalion Headquarters to return to Aden by June 20th. But on attempting to land at Khormaksar Airport at Aden the runway came under fire and they were diverted to Djibouti in French Somaliland, and to French hospitality. The resident Paras could hardly forbear reference to the peace and orderliness prevailing there in contrast with the rioting in Aden.

By the time Battalion Headquarters had finally managed to get back to Aden after their sojourn in Somaliland one company had already been committed to help in Sheik Othman, under command of the 1st Battalion the Parachute Regiment. They were to be there for six weeks, when conditions were at their most unpleasant, and there is little doubt that Sheik Othman was a good deal 'hotter' than the more publicized area of Crater. The remainder of the battalion were required for the task of protecting Little Aden and its vital refinery.

In August the battalion returned to England, and to Public Duties, after a memorable nine months' tour. An important one in the lives and experiences of all who were in Aden at that time, and a landmark in the history of the Regiment. Nine Irish Guardsmen lost their lives during this campaign, a campaign which showed that the traditions of the Irish Guards were in as good hands as ever, and proved that the modern Irish Guardsman was as adaptable to the changing pattern of war as had been his forebears.

The battalion were awarded one Military Cross, one Military Medal, two Mentions in Dispatches, two Queen's Commendations, fourteen Commander-in-Chief's Commendations and two South Arabian Army Special Commendations during their tour in Aden.

Postscript

On October 1st, 1967, the 1st Battalion Irish Guards mounted Queen's Guard – the wheel had turned full circle once again. Peacetime soldiering was then to be their lot until November 1970 and the departure for Hong Kong and the Regiment's first tour to the Far East. The feature of soldiering in the interim was the overseas training which took the Irish Guards to Canada in 1968 and to Cyprus the following year.

A sad event for all Irishmen was the merging in 1969 of the three Irish infantry regiments, the Royal Inniskilling Fusiliers, the Royal Irish Fusiliers and the Royal Ulster Rifles, into a new regiment known as the Royal Irish Rangers. But overshadowing everything for the Irish Guards was the death of their Colonel, Field-Marshal Earl Alexander of Tunis and Errigal, KG, GCB, OM, GCMG, CSI, DSO, MC, DCL, LLD, PC, on June 16th, 1969. There were then no men present in the battalion, who had actually served under Alex, but from the newest-joined Mick straight out of the Guards Depot, who had probably only once or twice caught a glimpse of the great man, to those who had known and revered him all their lives, it was a great personal loss. The huge number of his old comrades who attended the Memorial Services to the Field-Marshal in Ireland and in England was testimony enough of the regard in which they held the 'Greatest Mick of all'. There could have been no warmer compliment to a great Irishman and a great Irish Guardsman. Lieutenant-General Sir Basil Eugster, KCVO, CB, CBE, DSO, MC, was appointed Colonel of the Regiment.

Over the past months history has repeated itself in bitter fashion, and once again Ireland is in torment – although only isolated parts of Ulster are affected. The 1st Battalion Irish Guards have not been involved in the difficult task of maintaining law and order but like all Irishmen – and the battalion is still currently eighty per cent Irish – they have been shocked and appalled by the excesses in Northern Ireland. Yet once again it is good to report that, as in 1921, the Irish Guards, and the other Irish Regiments in the British Army, have remained aloof from the turmoil in their country. And it might be relevant to quote an extract from an article in the *Daily Sketch* written as long ago as 1944: '...And here, surely, is a lesson for all of us in Great Britain and anywhere else where ancient feuds survive in all their original vigour and, I am afraid I must add to many of them, in all their original stupidity. For what has happened in the case of the Irish Guards demonstrates how easily, if the spirit is willing, long-standing quarrels can be composed without any pressure from above or below...'

It has been a fascinating experience for me in the preparation of this book to watch the evolution of the present-day Irish Guardsman, to see the parallels and the contrasts with his predecessors. There have been changes, of course, and the modern soldier, and not least the Irish Guardsman, is more reliant on his comforts, and man-for-man he is probably less self-sufficient and mentally tough than his forebears; on the other hand he is a more highly trained, sophisticated individual fighting man able to use a bewildering variety of weapons and an expert in the employment of a spectrum of techniques.

But beneath the sophistication of the modern soldier can still be seen the Mick of old; his attitude towards his superiors, his natural politeness and courtesy, the sense of occasion are as they have always been with Irish troops. His exuberance in battle and relish of a good fight are characteristics as valid today as they were in the age of the Wild Geese.

For the Irish Guards, though, the most dominant feature is that their spirit is completely unchanged. It is this spirit which has sustained the Regiment for the past seventy years and which will undoubtedly do so for many years to come. It was Kipling who wrote of the Irish Guards, 'Their discipline is that of the Guards'; and he could have added 'but there is a touch of magic about all they do – this is theirs alone'.

QUIS SEPARABIT?

The Guards Division

On the formation of the Irish Guards in 1900 the Brigade of Guards as it then was – the 'Brigade' as it was always known – comprised three Regiments of Foot Guards, the Grenadier Guards, the Coldstream Guards and the Scots Guards; and in existence then were three Regiments of Household Cavalry, the 1st and 2nd Life Guards and the Royal Horse Guards (the Blues).

In overall command of the Sovereign's Household troops is, to give him his full title, the Major-General Commanding the Household Division and the General Officer Commanding the London District, who is known as 'the Major-General'.

To preside over their separate interests each Regiment of Foot Guards has a Regimental Headquarters whose function it is to supervise such matters as recruiting, officer selection and other aspects of general policy on behalf of the whole Regiment. The Service Battalions come under their jurisdiction for all matters concerning Regimental policy wherever they are serving, and when they are in London come under command for all purposes except local administration. Commanding each Regiment is the Regimental Lieutenant-Colonel, a full Colonel's appointment, and to assist him is a Regimental Adjutant with a small staff.

Another function of the Regimental Headquarters is to control the Regimental Bands which are such a feature of ceremonial life in London and on tour. The Regimental Bands are a completely separate almost non-military entity composed exclusively of professional and high-class musicians. The members of the bands are in fact musicians first and soldiers incidentally; almost all have off-duty employment in orchestras or bands.

To provide music for the service battalions each has a Corps of Drums, the Scots and Irish Guards having Pipes in addition. The members of the Pipes and Drums are an integral part of the 1st Battalion Irish Guards and, in war, man the Intelligence Section, act as stretcher-bearers or form a fighting platoon.

The Guards Depot, which is now at Pirbright Camp near Woking in Surrey, is the home and training ground for continuing generations of Guardsmen. Here the recruits are taught the innumerable military arts which the modern soldier must know and are indoctrinated with the discipline and standards of the Guards – standards which have never wavered over the past three hundred and more years of the oldest Regiments of the Foot Guards.

To the public the peacetime ceremonial is the most conspicuous role of the Foot Guards. Observing the intricate ceremonial of Guard Changing, Guards of Honour and the crowning spectacle, the Queen's Birthday Parade, it is difficult for the layman to appreciate that the men taking part in the pomp and trappings of seemingly stereotype ceremonial drill are some of the finest and most versatile fighting troops in the world. For the Regiments of Foot Guards have fought in most of the major campaigns in British history, and a list of their battle honours is a roll-call of military prowess over the past three centuries.

Nowadays the Brigade of Guards, which since 1968 should strictly be called the 'Division', for the Household Brigade changed its title to Household Division in that year, consists of two battalions of Grenadier Guards, two battalions of Coldstream Guards, two battalions of Scots Guards, and one battalion each of the Irish and Welsh Guards. The Household Cavalry comprises the Life Guards, and a recent amalgamation of the Royal Horse Guards (The Blues) and The Royal Dragoons (1st Dragoons), which is known as The Blues and Royals. In addition there are an independent Guards Parachute Company and a Guards Company of the Special Air Service Regiment.

Index